Life of Fred®

Kidneys

Life of Fred

Kidneys

Stanley F. Schmidt, Ph.D.

Polka Dot Publishing

© 2015 Stanley F. Schmidt
All rights reserved.

ISBN: 978-1-937032-06-7

Library of Congress Catalog Number: 2012912474
Printed and bound in the United States of America

Polka Dot Publishing   Reno, Nevada

To order copies of books in the Life of Fred series,

visit our website PolkaDotPublishing.com

Questions or comments? Email the author at lifeoffred@yahoo.com

Fourth printing

Life of Fred: Kidneys was illustrated by the author with additional clip art furnished under license from Nova Development Corporation, which holds the copyright to that art.

for Goodness' sake

or as J.S. Bach—who was never noted for his plain English—often expressed it:

Ad Majorem Dei Gloriam
(to the greater glory of God)

If you happen to spot an error that the author, the publisher, and the printer missed, please let us know with an email to: lifeoffred@yahoo.com

As a reward, we'll email back to you a list of all the corrections that readers have reported.

A Note Before We Begin
Life of Fred: Kidneys

It is not possible to have a mathematics book entitled *Kidneys*. Or worse yet, *Life of Fred: Kidneys*. What will people think?

But I can't help it. Other authors might write books with snappy titles such as:

MATH FOR GRADE 4
ARITHMETIC EXERCISES
WORKBOOK FOR MULTIPLICATION
MORE DRILL WORK FOR YOUR KID
POUND IT IN
GRADE THREE NUMBERS
PAGES OF PROBLEMS

. . . but I can't inflict that kind of pain on children. It just isn't in me.

> Children will learn more from three problems done with joy than from thirty drill-and-kill exercises.

Here is a picture from my childhood. Two are serious. Two are smiling. And one wrote *Life of Fred*.

I haven't been able to stop laughing. Life is filled with too much joy.

the author ↑
of *Life of Fred*

But *Life of Fred: Kidneys* is . . .

NOT JUST FUN AND GAMES

There is a lot of mathematics in this book. For example, we will discuss areas and volumes, graphing, the domain and codomain of functions, one of the most common errors in statistical thinking, elapsed time, perimeters, graphing, the meaning of $\frac{1}{6}$, the logical equivalence of P → Q and not-Q → not-P, exponents, and why $\sum_{i=1}^{52} i$ equals 1,378.

In the last *Your Turn to Play* in this book, is the question: "How many polka dots can fit in a glass?" That is not just a silly question. It is a two-step problem that asks the reader to first find the last term of an arithmetic series and then find the sum of the series.

HOW THIS BOOK IS ORGANIZED

Each chapter is about six pages. At the end of each chapter is a *Your Turn to Play*.

Have a paper and pencil handy before you sit down to read.

Each *Your Turn to Play* consists of about four questions. Write out the answers—don't just answer them orally.

After all the questions are answered, take a peek at my answers that are given on the page following the questions. At this point you have *earned* the right to go on to the next chapter.

Don't just read the questions and look at the answers. You won't learn as much taking that shortcut.

CALCULATORS?

Not now. There will be plenty of time later (when you hit Pre-Algebra). Right now in arithmetic, our job is to learn the addition and multiplication facts by heart.

Contents

Chapter 1	Keys to Success.	13
	talent, practice, patience	
	numerals	
	the fraction $\frac{1}{4}$	
	borrowing in subtraction	
Chapter 2	Administration Building.	19
	functions, domains, codomains	
	what teachers' salaries are based on	
	using an ATM	
	how to write a check	
Chapter 3	Police Station.	25
	idioms	
	whole notes, half notes, quarter notes	
	elapsed time computations	
	in libraries, *McCoy* comes before *MacDowell*	
Chapter 4	Finding Coalback.	31
	two out of a hundred = 2%	
	staying on the subject when talking	
	the safe way to catch crooks	
	in libraries, *20,000 Leagues Under the Sea* comes before *The Unknown Night*	
Chapter 5	To Catch a Thief.	37
	brothers protect their sisters	
	what 100% means	
	reading a clock	
	table salt = NaCl	
	converting ounces to grams	
Chapter 6	Functions.	43
	using arrows to describe a function	
	images of elements of the domain	
	playing guess-the-function	
	computing how many great-great-great-...-great grandparents you had 30 generations ago	

Chapter 7	Jelly Beans and Sugar Weasels. 49	

        volume of 4' x 2' x 3' hole in the ground
        finding one-fourth of something
        converting 115 ounces to pounds (dividing with a remainder)

Chapter 8   Errors in Thinking. 55

        does "12 noon" make sense?
        the biggest error in thinking
        education and life-time earnings
        what to do when you hit a wall in life
        computing a perimeter
        answering a problem in English vs. scribbling down some numbers

Chapter 9   The President's House. 61

        different functions with the same domain and codomain
        giving house tours to make a million dollars
        one-sixth of a cracker
        the only new thing in calculus: the concept of *limit*
        filling in missing dimensions on a map

Chapter 10   The Tour.. 67

        a stove with 24 burners
        a basic law of economics
        area of a 37-foot x 48-foot bedsheet
        the six ways to pay $15

Chapter 11   The Rooms. 73

        a score, a dozen, a brace
        polar form of complex numbers
        solving $x^5 = 32$
        happiness and owning polar bears
        your becoming a university president

Chapter 12   Nine Days.. 79

        how to carry a polar bear doll
        a quarter after two = 2:15
        what to pack for camp
        adding up the costs for camp

Chapter 13   To See Dr. Morningstar. 85

        Anglo-American Cataloguing Rules
        stethoscopes
        several different meanings of *specimen*
        what floor room 765 is on

Chapter 14   Leisure. 91
                 thinking about the bigger questions in life
                 books—one place to find answers
                 why we have two kidneys
                 how to find one-half of something

Chapter 15   Packing. 97
                 parts of speech (nouns, pronouns, verbs, etc.)
                 a situation in which a lunch box with a duck on it is the right choice
                 $\{1, 2, 3\} \cup \{3, 4\} = \{1, 2, 3, 4\}$
                 $5x^2y + 2x^2y = 7x^2y$
                 point, segment, square, cube, and tesseract
                 twelve different approaches to set theory
                 the human brain was not designed to multi-task
                 the best time in history to become a mathematician
                 logically equivalent statements

Chapter 16   Spurs. 103
                 an example of adumbration
                 arithmetic sequences (first term, last term, common difference, number of terms)
                 $l = a + (n - 1)d$

Chapter 17   Miss Ente. 107
                 arithmetic series
                 the sum $s = (½)n(a + l)$
                 things are not always exactly how they seem
                 why Miss Ente hasn't married
                 shoe sizes—U.S., Euro, and U.K.
                 the countries of the United Kingdom
                 the countries of Great Britain
                 finding the sum of the first 67 terms of $8 + 14 + 20 + 26 + 32 + \ldots$

Chapter 18   Stacking Books. 113
                 how to make boots fit better
                 2 stacks of 6 books, 3 stacks of 4 books, etc.
                 why horses can't wave at you while they are galloping
                 four ways to talk: entertain, ask, convince, and inform
                 the sum of the first million terms of $43 + 44 + 45 + 46 + 47 + 48 + \ldots$
                 the Fundamental Theorem of Calculus

Chapter 19   Discovering Something New...................... 119
                      you are not defined by the things you own
                      sigma notation
                      how new mathematics is sometimes created
                      analytic geometry
                      how to add up:

$$\begin{matrix}\bullet\\ \bullet\bullet\bullet\bullet\\ \bullet\bullet\bullet\bullet\bullet\bullet\bullet\\ \bullet\bullet\bullet\bullet\bullet\bullet\bullet\bullet\bullet\bullet\\ \bullet\bullet\bullet\bullet\bullet\bullet\bullet\bullet\bullet\bullet\bullet\bullet\bullet\\ \bullet\bullet\bullet\bullet\bullet\bullet\bullet\bullet\bullet\bullet\bullet\bullet\bullet\bullet\bullet\bullet\\ \bullet\bullet\bullet\bullet\bullet\bullet\bullet\bullet\bullet\bullet\bullet\bullet\bullet\bullet\bullet\bullet\bullet\bullet\bullet\\ \bullet\\ \bullet\end{matrix}$$

                      from 3:45 to 4:10
                      summer solstice

Index. 125

Chapter One
Keys to Success

Fred is five years old. He has been a professor at KITTENS University for years. He and his doll, Kingie, have an office on the third floor of the Math Building.

Kingie was getting ready to do his first painting of the day. He squeezed out dabs of oil paint onto his palette and stared at the blank canvas.

It was the first day of the month and he wanted to start a new series of paintings. He had just finished 72 paintings of deer. His favorite was "Fawn in Forest."

"Fawn in Forest" by Kingie

Kingie had a natural talent for painting. Fred was less talented.

"Small Deer" by Fred

Kingie also has two other advantages over Fred. He has spent years practicing painting, and he takes his time. Kingie spent 40 minutes painting "Fawn in Forest."

Fred spent 15 seconds drawing "Small Deer." Fred did not have much patience when it came to doing art.

- ★ Talent
- ★ Practice
- ★ Patience

} Success!

Chapter One Keys to Success

Not Kingie's game

Years ago, Kingie realized that he didn't have any natural talent for playing soccer. He is a beanbag doll and doesn't have legs.

Kingie told Fred, "I have decided what my next series of paintings will be. I am going to paint numbers. I'm going to start with the numeral 3."

Kingie got out a piece of paper and tried out various ideas. He was patient and didn't just start painting.

$3 \quad \mathbf{3} \quad \mathit{3} \quad \boxed{\cdot\,\cdot\,\cdot} \quad 3$

"I bet no other artist has ever painted numbers before," Kingie said.*

Fred also liked the number three. It was one of his favorite numbers. He also liked 0, 1, 2, 4, 5, 6, 7, 8, 9, 10, 11, 12, 13. . . . He liked fractions such as $\frac{1}{4}$.

He liked decimals. He recently bought a bow tie for four dollars and twenty-six cents. ($4.26) The dot between the 4 and the 2 is called a decimal point.

---

* Kingie was wrong.
   In 1928, Charles Demuth painted the numeral 5. He called it "I Saw the Figure 5 in Gold." That painting is now in the Metropolitan Museum of Art in New York.

14

Chapter One — Keys to Success

Fred liked to wear a bow tie when he taught. He thought that it made him look older. If he would remember to remove the price tag, he would look a little less silly.

"It's the first of the month," Kingie told Fred. "Don't forget to pick up your paycheck."

"Oh. I had forgotten about that," Fred said. "Thank you for reminding me."

"You're welcome. I'm glad to be of help."

Fred was sitting at his desk in his office preparing to teach his classes.

He had a little card on his desk to remind him of his schedule. One of the big joys in his life was teaching mathematics. Even if KITTENS University only paid him $10 a month, he would still be very happy.

```
8–9    Beginning Algebra
9–10   Advanced Algebra
10–11  Geometry
11–noon Trigonometry
noon–1 Calculus
1–2    Statistics
2–3    Linear Algebra
3–3:05 Break
3:05–5 Seminar in Biology,
       Economics, Physics, Set Theory,
       Topology, and Metamathematics
```

✓ He slept in his office so he didn't need to pay for housing.

✓ He spent almost nothing on food. If he did buy something to eat, he would usually put it in his desk drawer "for later."

✓ Clothes? He hadn't grown an inch in years, so all his clothing still fit. The only thing in a clothing store that excited him was the bow tie department. Fred owned some unusual bow ties.

feet bow tie   harp bow tie

Chapter One   Keys to Success

✓ Cars? He was a little too young to drive.

✓ There was one thing Fred found irresistible. He loved books. Every wall in his office had bookshelves, and every bookshelf was filled. There were stacks of books on the floor.

Fred figured that if he lived to the age of 80, he only had 75 years left to read. That seemed like a very short time to finish all the books he wanted to read.

$$\begin{array}{r} \overset{7}{\cancel{8}}\overset{}{\cancel{0}} \\ -5 \\ \hline 75 \end{array}$$

Here's the language that goes with that subtraction problem: You can't subtract 5 from 0. You borrow 1 from the 8. That turns the 8 into a 7. The 0 now becomes 10. Five from 10 is 5. Nothing from 7 leaves 7.

Besides the books that he owned, Fred would often go to the KITTENS University Library and to the public library and borrow armfuls of books.

It was 7:30 a.m. He had been working at his desk for two hours. His eight o'clock beginning algebra class would start in a half an hour. There was just enough time to pick up his paycheck, deposit it in the bank, and get to class.

He hopped off his chair, said goodbye to Kingie, and headed out his office door.

When Kingie was not painting, Fred would give him a hug before he left. When he was painting, Fred just said goodbye. Hugging a doll that is holding a palette full of wet oil paint can be a big mistake.

Fred headed down the hallway past the nine vending machines, down two flights of stairs, and out into the cool morning air. He suddenly realized that he hadn't gone jogging this morning, so he broke into a run to get some exercise. He ran past the tennis courts, the university chapel, and the rose garden. Just north of the rose garden was the Administration Building.

Please get out a piece of paper and write out the answers to each of these *Your Turn to Play* questions before you look at the answers on the next page. You will learn more if you do it this way.

Your Turn to Play

1. If bow ties cost $4.26 apiece, how much would two bow ties cost?

2. When Kingie put dabs of oil paint on his palette, he would squeeze out 9 ml (milliliters) from each tube of paint. If he used 23 colors, how many ml would be on his pallette?

3. If it took Fred two hours to prepare for his classes and he finished his preparation at 7:30 a.m., when did he begin?

4. Fred made seven drawings of baby animals.

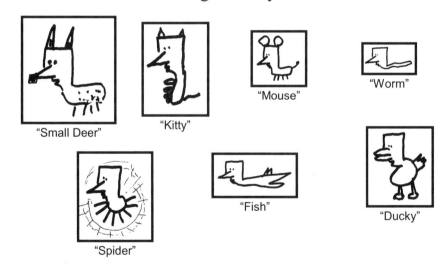

If it took him 15 seconds to make each picture, how long did it take him to do all 7 pictures?

17

······· COMPLETE SOLUTIONS ·······

1. There are two ways you might have done this problem.

By addition: $4.26
              + $4.26
              ─────
               $8.52

By multiplication: $4.26
                   ×   2
                   ─────
                    $8.52

2. A milliliter (ml) is the same as a cubic centimeter. One milliliter is the volume of a box that is one centimeter on each side. A centimeter is about the width of your baby finger.

We need to find 9 times 23.

Hard way: 9
          × 23
          ────
            27
          18
          ────
          207 ml of paint

Easier way: ²23
            × 9
            ───
            207

3. Two hours earlier than 7:30 a.m. is 5:30 a.m.

4. There are two ways to find out how long it would take to make 7 drawings when each drawing takes 15 seconds:

Hard way: 15
            15
            15
            15
            15
            15
          + 15
          ────
          105 seconds

Easier way: ³15
            × 7
            ───
            105

18

Chapter Two
Administration Building

Fred had to go to the Administration Building to pick up his check. It was the largest building on the KITTENS campus. Fred could never understand why the building that kept student records, sent out letters, and handed out paychecks was so much more impressive than the rest of the buildings on campus.

Fred thought to himself *If the job of a university is to educate students, shouldn't the classrooms be the most important places on campus?*

Administration Building

Math Building where Fred's office is

In the Administration Building, there were gold doorknobs on every door. In the hallways there were fine oil paintings and statues of all the former university presidents.

In the Math Building, some of the doorknobs were falling off. Instead of statues, there were vending machines.

He entered the Administration Building. The floors were marble and the ceiling was 18 feet tall. Fred could not imagine why the ceilings needed to be so tall. He thought *I'm only three feet tall. You could stack six of me to reach the ceiling.*

$$\begin{array}{r} 6 \\ 3\overline{)18} \\ \underline{18} \\ 0 \end{array}$$

Chapter Two   Administration Building

Fred stood in line with other teachers who were waiting to pick up their paychecks.

The 52-year-old received $5,200.
The 37-year-old received $3,700.
The 84-year-old received $8,400.
The 28-year-old received $2,800.
The 24-year-old received $2,400.

And Fred, who is 5 years old, received $500.

At KITTENS University, the salary of a teacher was a function of the age of the teacher.*

For this function, the domain is the age of the teacher and the codomain is the salary. The function is the rule that associates to each age a salary.

It's not too hard to guess that a 45-year-old would be paid $4,500.

---

* This is *not* the normal way that teachers are paid. More commonly, a salary depends on such things as:
- how much education the teacher has
- how many years the teacher has taught
- whether the teacher is an instructor, an assistant professor, an associate professor, or a full professor.

Chapter Two Administration Building

Fred raced off to the bank to deposit his check. At the bank he stood in line at the ATM (automated-teller machine). The people in the line looked familiar.

When he got to the ATM, he often asked someone to pick him up so that he could use the machine. When no one else was around, Fred would go into the bank and bring a chair outside to stand on.

He deposited his check and withdrew $40 in cash. He now had $460 in his checking account.

$$\begin{array}{r} \overset{4}{\$\cancel{5}}00 \\ -40 \\ \hline 460 \end{array}$$

You can't subtract 4 from 0. You borrow 1 from the 5. That turns 5 into 4. The 0 now becomes 10. Four from 10 is 6. Nothing from 4 is 4.

The ATM gave him two $20 bills.

$$\begin{array}{r} 20 \\ +20 \\ \hline 40 \end{array} \quad \text{or if you prefer}\ldots \quad \begin{array}{r} 20 \\ \times2 \\ \hline 40 \end{array}$$

Last week Fred had borrowed $40 from Kingie. Now that he had some cash, he wanted to repay the debt. He had just

21

enough time to get back to the Math Building, repay Kingie, and get to his eight o'clock class.

Fred walked down the street carrying the two twenty-dollar bills in his hand. He was thinking that he could quickly give the money to Kingie if he didn't put it into his wallet or his pocket.

Fred had never noticed that people do not normally carry twenty-dollar bills in their hand as they walk down the street.

Someone bumped him. Bumped him hard. Grabbed his money and ran. It was someone who lived illegally on the third floor of the police station. He lived there with his sister. We won't mention any names.

Fred got up. He was a little dazed* but he wasn't hurt. He didn't know what to do. Clef, the music professor at KITTENS, ran over to Fred and said:

> Are you okay? He hit you pretty hard.
>
> It's a shame that there are robbers like that. They don't care about anyone except themselves.
>
> I got a good look at him. It was C. C. Coalback. If it weren't for him, there would be very little crime around here.

---

* dazed = stunned, shocked.

Chapter Two   Administration Building

Clef suggested that they go to the police station to report the robbery.

Fred nodded and off they went. Fred held Clef's hand.

Your Turn to Play

1. Fred weighs 37 pounds. Coalback weighs 243 pounds. How much heavier is Coalback than Fred?

2. Fred was glad that he had only taken $40 out of the ATM. If he had taken out $100 instead of $40, how many twenty-dollar bills would he have been carrying?

3. Can you think of two reasons why Coalback picked on Fred instead of Clef?

4. As they walked to the police station, Fred decided that instead of paying Kingie in cash, he would write a check. What two things did Fred forget in writing this check?

```
Fred Gauss                                          1357
Room 314, Math Building
KITTENS University              Date  June 1

Pay to the
order of   Kingie                       $ 40 00

Forty and 00/100                                 dollars

        Kittens Bank

"IT'S IN THE KITTY"
```

Chapter Two — Administration Building

·······COMPLETE SOLUTIONS·······

1. The General Rule when you don't know whether to add, subtract, multiply, or divide is to restate the problem using simple numbers. Then look at what you would do with the simple numbers. For example, if Fred weighed 2 pounds and Coalback weighed 7 pounds, then Coalback would be 5 pounds heavier than Fred. Did we add, subtract, multiply or divide? We subtracted 2 from 7.

So in the original problem, we subtract 37 from 243.

$$\begin{array}{r} 243 \\ -37 \\ \hline 206 \end{array}$$

Coalback is 206 pounds heavier than Fred.

2. How many twenty-dollar bills in $100? Using the General Rule, we might ask, How many five-dollar bills in $15? There are three of them. We divided $15 by 5.

So in the original problem:

$$20 \overline{\smash{\big)}\,100}$$ with quotient 5, $100 - 100 = 0$.

There are five twenty-dollar bills in $100.

3. The two reasons that I could think of are:

  1) Fred is a lot smaller than Clef, and

  2) Fred was carrying his money in his hand. It was visible.

4. Fred forgot to put the year in the date, and he forgot to sign the check.

Writing the Numerals in a Check

You write $40^{00}.

You don't write $40.00 (because that can be changed into $40,000.00 by turning the period into a comma.)

You don't write $ 40.00 (because that can be changed into $940.00 by inserting a number between the $ and the 4.)

Chapter Three
Police Station

Clef didn't know where the police station was. Fred told Clef that he had been there on several occasions. Clef didn't ask why a five-year-old had been to the police station that often.

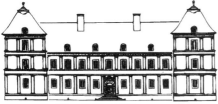
The Millard Fillmore
Federally Funded Police Station

Fred led Clef to the station. It was a huge three-story building, but it was smaller than the Administration Building.

The inside of the station had changed since Fred's last visit. They had added a new reflecting pool and a big white statue.

Fred wanted to go swimming, but Clef told him this was not the right time to do that.

A sign indicated that the police office was on the first floor in the back. There was only one officer in the entire building since there was very little crime near KITTENS University.

Chapter Three   Police Station

As they walked down the polished marble hallway, Clef spotted someone heading up the stairs. This was strange since the only occupant of the building was the police officer on the first floor.

"Well, I'll be a sforzando on a semibreve!* Do you see what I see? That's Coalback going up the stairs! What's he doing here?"

They rushed through the winding halls on the first floor to the police office. As they entered the office, the police officer said, "Hi, sonny. Do you have another postcard to show me?"

Fred was surprised. He had brought that postcard to the police officer back in February (in *Life of Fred: Dogs*). The police officer still remembered it. That really shouldn't have been so surprising. The police officer had done very little work in March, April, and May.

Clef interrupted, "We have a crime to report."

The police officer said, "I'm all ears."

> Time Out!
>
> *To be all ears* is an **idiom**. It means to be very attentive. It means to be really listening.
> It does not mean that the police officer was a pair of 60-pound ears with no legs or arms.
>
> In order to be a good reader, you need to learn that not everything you read is literally true.

---

* A semibreve is not a small swimsuit. Since Clef is a musician, he might say *semibreve* while most people call it a whole note. *Sforzando* is the only word in your dictionary that begins with s-f-o.

whole note   half   quarter   eighth

26

Chapter Three   Police Station

> Having *a chip on your shoulder* doesn't mean you are carrying around a piece of wood. It means you are looking for a fight.
>
> When you say that she is *the apple of my eye,* you don't mean that fruit is growing in your eyeballs. It means you cherish her above all other people.
>
> *Barking up the wrong tree* has nothing to do with dogs or trees. It means you are making a mistake in your approach to a problem.
>
> If you *bend over backwards*, it means you are really willing to help.
>
> *Blue moons* are events that occur rarely.
>
> *To buy a lemon* refers to getting a car that has lots of problems.
>
> *Crack someone up* means to make them laugh.
>
> This   and not this

Clef continued, "This young man is Professor Fred Gauss who teaches in the math department at KITTENS University. His office is on the third floor of the Math Building. He lives there with his doll, Kingie, who is a world-famous artist. Perhaps, you have seen some of Kingie's art. Last year I bought two of his oil paintings that now hang in my office. I love to look at them while I practice the violin. I often practice my violin for several hours each day because I'm preparing for a concert this fall in South Carolina. It is such a lovely state in the fall. Besides the gorgeous trees and flowers, the autumn sun falling on some of the old mansions is a sight you don't want to miss. I once had the opportunity to stay

Chapter Three Police Station

overnight at the MacDowell mansion in South Carolina. In the morning, they served me one of the most splendid breakfasts I have ever had. The lingonberry jam on scones was the best in the world. Some people call it mountain cranberry, but I prefer to call it lingonberry. *Lingon* is a Swedish word."

Fred was excited by all he was learning.

The police officer was going nuts (idiom). He interjected, "Could you please cut to the chase?*"

Fred said, "I got robbed."

"When did this happen?"

Clef began, "I recorded the exact time on my new watch that I purchased last week. It is a Swiss watch that is guaranteed accurate to within one minute every century. That means that a hundred years from now it might be either 8:23 a.m. or 8:24 a.m. or 8:25 a.m. Right now, of course, it is 8:24 a.m."

Fred thought to himself *It's already 8:24! I'm missing my eight o'clock beginning algebra class. I'm sure that Betty will teach it since I'm not there.*

Clef continued, "The Swiss are very fine watchmakers. They are also known for their excellent chocolate. If you have ever tasted it, you would know what I mean. I received a pound of it as a birthday gift last November."

"When did the robbery occur?" the police officer asked.

Fred answered, "A couple of minutes ago."

"Who robbed you?"

Clef was happy to relate, "I had a very clear view of the incident and can positively identify the malefactor [MAL-eh-fac-tor = the wrongdoer]. The light was very good and I just had my eyes checked at the optometrist two weeks ago. He told me that I had no trace of cataracts [clouding of the lens of the eyes] yet and gave me a new prescription. I got my new glasses last week, and they have been wonderful. I chose the black frames, as you can see, rather than the brown frames I used to have. I think they make me look more distinguished."

Fred said, "It was C. C. Coalback. We both saw him."

---

* *Cut to the chase* is an idiom. It means to eliminate all the unnecessary stuff and get to the point.

Chapter Three Police Station

Your Turn to Play

1. When the police officer asked a question, Clef might take 300 words to answer it. Fred might take 18 words. How many more words would Clef take than Fred?

2. If it is currently 8:24, which means 24 minutes past eight o'clock, what time was it five minutes ago?

3. Let's assume that the current year is 2012. Let's suppose that Kingie bought some oil paints from McCoy Art Supplies and sent them this check.

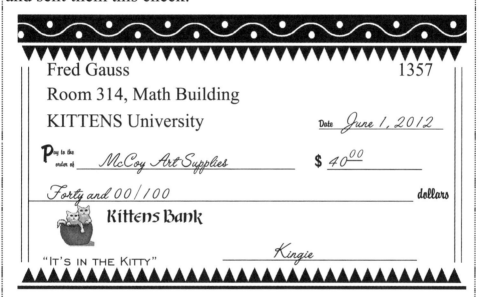

This is a very tricky question: What is wrong here?

Hints: ✓ He dated it correctly.
✓ He wrote the correct amount.
✓ He signed the check and didn't just print his name.
✓ He spelled *McCoy Art Supplies* correctly.

Chapter Three — Police Station

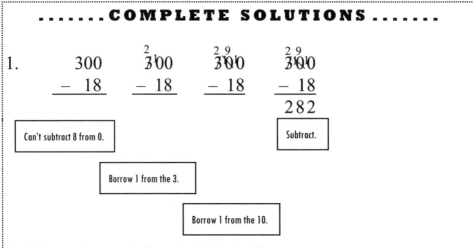

......COMPLETE SOLUTIONS.......

2. Five minutes before 8:24 is 8:19.
3. Kingie was using one of Fred's checks. That $40 would be coming out of Fred's checking account, not Kingie's.

Why Being a Librarian Can Be Hard

You have to alphabetize the titles of your books in the library catalog. This is more than just knowing your ABC's. Everyone knows that *accept* comes before *access*. That part is easy.

In the chapter you have just read, we talked about the MacDowell mansion and the McCoy Art Supplies company. If you were alphabetizing *MacDowell* and *McCoy* for a library catalog, which would come first?

Most libraries use the alphabetizing rules in the ANGLO-AMERICAN CATALOGUING RULES VERSION 2. According to those rules, *McCoy* comes before *MacDowell*!

The rules state that authors whose last names begin with *Mc* will be alphabetized as if their names were *Mac*.

The encyclopedia at my house also follows this rule. My dictionary does not.

Chapter Four
Finding Coalback

When Fred said that C. C. Coalback had robbed him, the police officer sighed. Sighing is the sound you make when you exhale (blow out) in weariness or sadness. Some people say that the wind sighs when it blows through the trees.

"That name is very familiar," the officer said. "If it weren't for Coalback and his sister, there would be almost no crime at KITTENS."

Fred and Clef were silent.

The officer continued, "I could leave my watch or a ten-dollar bill on my desk, and most people would never touch either of them. Most people are honest. But maybe two people out of a hundred would steal the moment that my back was turned."

Fred thought mathematically: two out of a hundred = two percent = 2%. That meant that 98% were not thieves.

"Coalback is a hard guy to catch," the officer said. "Back in February he and his sister escaped from jail, and we haven't been able to find them. There is a reward of $1,000 for anyone who can help in his capture."

Clef's eyes lit up. A thousand dollars could buy a lot of whole notes 𝅝, half notes 𝅗𝅥, quarter notes ♩, and eighth notes ♪.

He excitedly said . . .

> Time Out!
> Before we report what Clef said, I need to mention some things that have probably

> become pretty obvious to most readers.
>
> First of all, he likes to talk. This, by itself, is not a bad trait for a teacher to have. Can you imagine what it would be like to have a teacher who stood up in front of a classroom and never said anything?
>
> Second, Clef knows a lot of things. (Never write *alot*. It isn't a word.) He can talk about semibreves, South Carolina, and lingonberry jam. This is also a nice trait for a teacher.
>
> Third, it is obvious that Clef's mind wanders. He can't stay on a single subject. This makes him a very bad teacher. That is why his words are put in 9-point type. They are hardly worth reading.

Responding to the officer's statements that Coalback was hard to find and that there was a $1,000 reward, Clef said, "Just as we walked into this police station,* I told Fred that he shouldn't be swimming in the reflecting pool that is right next to the large white statue. Since the statue is white, I assume that it isn't granite, because granite is gray. *Gray* can be spelled in two different ways: g-r-a-y and g-r-e-y, with g-r-a-y being the preferred spelling in the United States. If you ask people what their favorite color is, very few will say gray, or black, or white. And yet, if you look at cars in a parking lot, there are lots of gray, black, and white cars. If eight-year-old girls were allowed to choose the color of the family car, I would bet that there would be many more pink cars and purple cars. The next car I buy will have pink and purple stripes on it and will have some music printed on the hood. That way I will be able to find it easily in a parking lot.

---

* At this point, if Clef had simply said that he saw Coalback head up the stairs, he would have received the thousand-dollar reward.

Chapter Four   Finding Coalback

Before Clef could continue with how he might buy a Japanese or a German car and that the Japanese and the Germans were the losers in World War II, Fred did something he normally didn't do.

He interrupted. "Coalback's upstairs," he said. Those two words earned Fred a thousand dollars.

"Upstairs!" the police officer exclaimed.

"Yes," Fred replied. "We saw him head up the stairs as we came into the building."

The policeman was in a state of shock. He sat down at his desk to give himself some time to think. Fred and Clef remained silent.

After a minute, the policeman knew that the first thing was to make sure that Fred and Clef were safe. He wrote a check for $1,000 and handed it to Fred. Then he told them to get out of the police station so that they wouldn't get hurt in case there was gunfire.

Fred and Clef did as they were told. They headed from the police office through the hallways to the exit. They only got lost once.

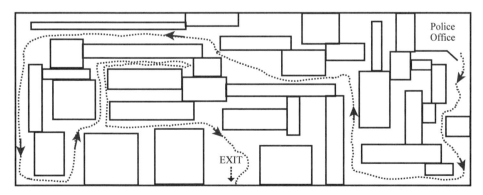

The police officer sat at his desk. He wasn't sure what to do next. He couldn't call for backup since he was the only one on the police force at KITTENS. He knew that Coalback and his sister were escaped convicts, but he didn't know how dangerous they were. One of the hardest things about being a police officer is not knowing whether the arrest will be easy or dangerous. The police officer did not want to be shot.

He had never been on the second and third floors of the police station. The federally funded Millard Fillmore Police Station was a giant waste of money. Only the officer's tiny office on the first floor had ever been used.

He looked through his file cabinet and found a map of the building.

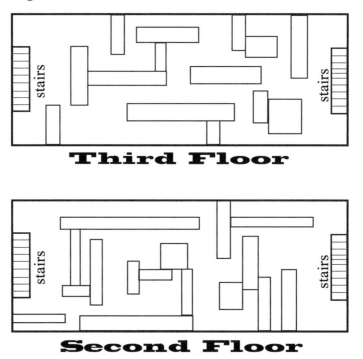

He saw that there were stairs on the left and right sides of the building. The map, of course, did not tell him where Coalback and his sister were hiding.

Rather than head up the stairs, the police officer decided to play it safe. He stood at the bottom of the stairs on the left

Chapter Four   Finding Coalback

side and yelled, "Okay Coalback! This is the police! You are under arrest! Come down with your hands in the air, and there won't be any trouble!"

Sometimes this approach works.

Your Turn to Play

1. Coalback and his sister were sitting in their room on the third floor of the building. The forty dollars that he had taken from Fred and three purses that he had snatched from women were on the table. Each purse had $30 in it.

How much money was on the table?

2. "Did you hear that?" Coalback's sister asked. "I think somebody was yelling at us to surrender."

"Don't be silly," Coalback said. "Nobody knows that we are up here."

When the police officer repeated, "Come down with your hands in the air," Coalback changed his mind. It took him 17 seconds to go through each purse and get the credit cards.

How long did it take him to go through all three purses?

3. While Coalback was getting the money and credit cards, his sister was sticking her two pearl necklaces into her pocket. One had 19 pearls on it and the other had 16 pearls. She had purloined [stolen] both of them from a jewelry store.

What was the total number of pearls?

> .... **ANSWERS** ....
>
> 1. Two ways to do this problem:
>
>    ```
>         40                    30
>         30         or        × 3
>         30                    90
>       + 30
>       $130                    90
>                             + 40
>                             $130
>    ```
>
> 2. ```
>         17
>        × 3
>        51 seconds
>    ```
>
> 3. ```
>         19
>       + 16
>        35 pearls
>    ```

Being a Librarian—Lesson 2

How would you alphabetize these two titles: 20,000 Leagues Under the Sea and The Unknown Night?

According to the ANGLO-AMERICAN CATALOGUING RULES VERSION 2, which most libraries use, 20,000 Leagues Under the Sea will be alphabetized as Twenty thousand Leagues Under the Sea.

Now which comes first: Twenty thousand Leagues Under the Sea or The Unknown Night?

Another rule from the AACR2 (as librarians like to call it) is that titles that begin with A, An, or The are to be alphabetized as if those articles didn't exist.

Final answer: Twenty thousand Leagues Under the Sea comes first.

I would rather be a mathematician.

Chapter Five
To Catch a Thief

Was this going to be an easy arrest or a dangerous one for the police officer? It was up to Coalback to decide. He could make it an easy arrest by simply putting up his hands and going down the stairs. Most crooks will do that. They don't want to risk getting tear gassed or shot.

In addition, if you resist arrest, that is another crime, and you will have to spend even longer in jail.

Coalback decided to sacrifice his sister.* First, he told her to give him the two pearl necklaces. He explained to her that when she was arrested, she shouldn't be carrying stolen property.

She handed them over to him.

Coalback knew that the police officer was at the bottom of the stairs on the left side of the building. He told his sister to walk down those stairs very slowly with her hands in the air. He told her to keep saying, "I'm coming. I'm coming."

She did that.

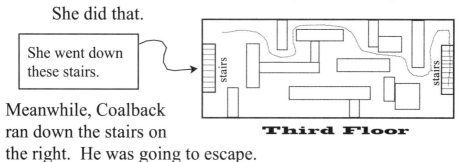

Meanwhile, Coalback ran down the stairs on the right. He was going to escape.

---

* Most good brothers would do almost anything to protect their sisters. If their sisters were in any kind of danger, they would do whatever was necessary to help them.
  Coalback loves only one person . . . himself.

Chapter Five   To Catch a Thief

When he got to the first floor, Coalback saw the open door to the police office. It was a perfect opportunity for him.

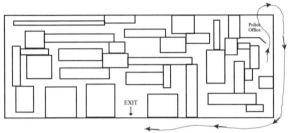

He knew that the police officer was on the other side of the building. A checkbook, a fountain pen, some sticks of gum, and a golf ball were on his desk.

He took them all and put them into his pocket. He crawled out through the window and headed around the outside of the building, saying to himself, "I am so clever. I am so clever."

He saw Fred standing there in front of the building with a thousand-dollar check in his hand. He failed to notice the music teacher, Professor B. Clef, who was standing next to Fred.

Clef had very strong hands from years of piano playing. He grabbed Coalback.

Coalback was helpless. He wiggled and squirmed but couldn't get free. Instead of exercising, Coalback had spent years watching television and eating junk food. He could be a bully with women whose purses he snatched or with 37-pound kids, but he was a marshmallow in Clef's grip.

Coalback begged Clef to let him go. "I will be good. I won't ever steal again. The checkbook, fountain pen, and these pearls accidentally got in my pocket. My sister made me do all the bad things that I've done. I'm innocent."

Fred couldn't believe what he was hearing. He knew that you were never supposed to lie. One hundred percent (100%) of what Coalback said were lies. It was all lies.

Chapter Five   To Catch a Thief

Clef started talking about fountain pens . . . how the first practical fountain pen was patented by Lewis Waterman in 1884 . . . how 1884 started on a Tuesday . . . and that was the year that the first Christian missionaries arrived in Korea. . . .

As Clef hauled Coalback to the police office, he gave a brief history of Korea, mentioned the effect of the Great Depression (which began in 1929) on Korean businesses, the unconstitutional programs Roosevelt started in the United States . . . and by the time they got to the police office, he was talking about the effectiveness of submarine warfare during World War II.

They arrived at the office at the same time as the police officer with Coalback's sister. Fred was still carrying the reward check in his hand.

The policeman said to Clef, "You actually caught Coalback. There's a $3,000 reward for making the capture. Let me write you a check before I forget."

He looked on his desktop for his checkbook. "I can't seem to find it. I left it right here. Is it okay if I send you the check later?"

Clef nodded. He and Fred left the police station and went out into the sunshine.

Clef looked at Fred and suggested that he put his $1,000 check in his pocket rather than carry it in his hand.

After they said goodbye to each other, Fred started walking to the bank to deposit the check. He thought about how Coalback was captured. He thought about paying Kingie the $40 he owed him. He thought about what he was going to do for the rest of his day.

Chapter Five   To Catch a Thief

Then he realized: I'VE GOT CLASSES TO TEACH!

He ran to the Archimedes Building where his classes are held. He looked at the clock on the wall. Since his beginning algebra class ended at 9 o'clock, he had 25 minutes left.

Betty was standing in front of the class writing on the board. She was one of Fred's oldest and best friends. When Fred was hired to teach at KITTENS University, he was only nine months old.* She used to carry him to class since he wasn't very good at walking yet.

Over the years she had learned a lot of math from Fred. She was now a graduate student.** When Fred didn't show up for class at 8 a.m., the students knew what to do (since Fred had told them at the beginning of the semester). They called Betty on her cell phone and asked her to come and substitute for him.

"Are you okay?" Betty asked. She was concerned. Fred very rarely missed class.

If Fred were Professor B. Clef, Betty would have received a twenty-minute talk about getting a salary . . . about how the word *salary* is related to the Latin word *salarium* where *sal* means salt . . . about how Roman soldiers were sometimes paid in salt . . . about the expression "being worth your salt" . . . about how in chemistry you can combine the metal sodium (Na) with the heavy greenish-yellow gas chlorine (Cl) and get ordinary table salt (NaCl).

---

* The whole story is in *Life of Fred: Calculus.* Fred had a very different childhood than most kids.

** Translation: She had graduated from college. She had her bachelor's degree. Graduate students are students who are working on an advanced degree, which is usually either a master's degree or a doctorate.

40

Chapter Five — To Catch a Thief

Instead, Fred said, "I'm fine."
Fred was eager to start teaching.

Some people might say that Fred was anxious to teach. Since the middle of the 1700s, the word anxious has had the second meaning of "desirous or eager." The primary meaning of anxious is related to anxiety, which means "full of mental distress."

If you say that you are "anxious to see your new grandson," does that mean that you are eager to see him, or does it mean that you are dreading seeing him because he might be really ugly? Anxious has both meanings. Eager doesn't.

Your Turn to Play

1. According to this clock, how many minutes after twelve is it?

2. The fountain pen that Coalback stole weighed 3 ounces. How many grams is that? (There are 28 grams in an ounce.*)

3. The 35 pearls that Coalback had in his pocket each weighed 7 grams. What was the total weight of the pearls?

4. Copy this on your paper and work it out: 58
                                             × 67

---

* Actually, it would be more accurate to say that there are 28.349527 grams in an ounce, but the only decimal points that you have had so far in your education have been in money. $8.39 means eight dollars and thirty-nine cents.

Chapter Five   To Catch a Thief

........COMPLETE SOLUTIONS........

1.  It is 25 minutes after twelve.

2. The General Rule when you don't know whether to add, subtract, multiply, or divide is to restate the problem using simple numbers. Then look at what you would do with the simple numbers. For example, if a fountain pen weighed 3 ounces and there were 4 grams in a ounce, then a fountain pen would weigh 12 grams. We multiplied.

So we multiply 3 times 28.
```
    28
  ×  3
    84
```

A fountain pen would weigh 84 grams.

3. By the same reasoning, we multiply 35 times 7.
```
    35
  ×  7
   245
```

The 35 pearls would weigh 245 grams.

4.
```
    58
  × 67
   406
   348
  3886
```

Chapter Six
Functions

Fred looked at the diagram that Betty had drawn on the board, and he knew immediately that she was in Chapter 11 of the *Life of Fred: Beginning Algebra Expanded Edition* story.

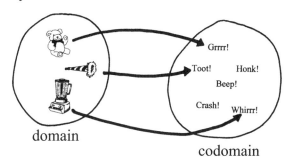

It was the story about a soldier named Thud. He put three items in his backpack: his little bear, a party horn, and a blender. That was the first set, which is called the domain.

The second set is the set of all sounds.

A function is a rule that associates to each member of the first set exactly one member of the second set.

A function isn't the first set.
 A function isn't the second set.
  A function is the rule that connects the two sets.

To the little bear, Thud associated Grrrr!
To the party horn, Thud associated Toot!
To the blender, he associated Whirrr!

It is the arrows that are the function. In order for it to be a function there has to be exactly one arrow coming from each member of the domain. Not two arrows. Not 15 arrows. Not zero arrows.

If you can count up to one, you can tell whether some rule is really a function. We want one arrow coming out of each member of the first set. We want each member of the first set to have one answer in the second set.

Fred drew some pictures of baby animals on the board. He called that the domain. The second set was some numbers. Then he drew some arrows

.

The class knew that Fred drew like a five-year-old. He had drawn a duck, a mouse, and a kitty. If Betty were at the board, her kitty might have looked like:

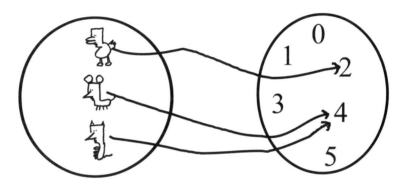

He asked the class if this was a function.

Joe, who was sitting in the back of the class eating jelly beans, held up his hand. Fred called on him.

Chapter Six   Functions

"I have a question," Joe announced. "Why does your ducky have both a bill and a mouth?"

Fred had never really looked at ducks very closely and thought that was the way a duck ought to look.

Fred repeated his question, "Is this a function?"

Chris, who was sitting in the front row, said that it was a function. There was exactly one arrow coming out of each animal in the domain.

⬥ The image of ducky is 2.
⬥ The image of mouse is 4.
⬥ The image of kitty is 4.

Each element of the domain has exactly one image in the codomain. That is the definition of a function.

Joe swallowed a bunch of jelly beans whole and said, "But, but, but the image of mouse and the image of kitty are the same!"

"Who cares!" exclaimed Chris. "Each element of the domain has to have one answer—one image—in the codomain. That's the definition. If all three arrows went to 5, it would still be a function."

Chris had a lot less patience with Joe than Fred had. That is one reason why Fred is such a good teacher.

"Now let's play the game of Guess-A-Function," Fred said. "I had a particular rule in mind when I drew those arrows. Can anyone guess what the image of a worm would be if I put that into the domain?"

Pat, who was sitting in the second row, knew the answer: "The image of the worm would be 0. The rule you were using is: *Count the legs.*"

Chapter Six   Functions

Joe put a whole handful of jelly beans into his mouth. One fell on the floor, and he smashed it with his foot. Custodians who clean the classrooms every night can always tell where Joe had been sitting.

"Functions are everywhere in mathematics," Fred said. "Can you guess what the rule is for this function?" He wrote on the board:

$$2 \rightarrow 6$$
$$5 \rightarrow 15$$
$$3 \rightarrow 9$$
$$20 \rightarrow 60$$

Almost everyone in the class guessed that the rule was: *Take the number in the domain and triple it.* Joe hadn't noticed that Fred had written that on the board. He was busy shoving the smashed jelly bean around with his foot.

"Here's a harder one. Here are ten examples."

46

Chapter Six   Functions

The domain is a set containing animals and humans. The codomain is {A, H}. The function is the rule: *Assign each animal to A, and each human to H.*

Your Turn to Play

1. Is this a function?

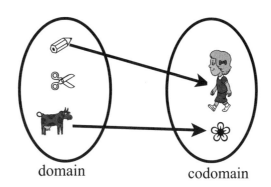

domain   codomain

2. This is a function. Its domain is pairs of numbers. You have studied this function in previous *Life of Fred* books. What is the name of this function?

        (3, 9) → 12   12 is the image of (3, 9) in this function.
        (7, 8) → 15
        (2, 2) → 4
        (10, 10) → 20
        (0, 8) → 8
        (200, 6) → 206

3. Multiplication is a function. The domain is pairs of numbers. If you are given a pair of numbers, there is always exactly one answer. What is the image of (49, 86)? Please don't use a calculator!

4. Why is this *not* a function? Let the domain be all living humans who are at least 25 years old. Let the rule be: *Assign each member of the domain to their oldest child.*

Chapter Six   Functions

······· COMPLETE SOLUTIONS ·······

1. It is not a function. Each element of the domain must be assigned to exactly one element of the codomain. ✂ was not assigned to anything in the codomain. It has no image in the codomain.

2. This is one of the first functions that you learned in math. It is called addition.

3.
```
    49
  × 86
   294
   392
  4214
```
(49, 86) → 4,214

4. In order for a rule to be a function it must assign to each member of the domain exactly one member of the codomain.

There are people who are 25 years old who have not gotten married and had kids. They are members of the domain but they would not be assigned to anything in the codomain under the rule: *Assign each member of the domain to their oldest child.*

One generation ago you have two parents. Two generations ago you have four grandparents. Three generations ago you have eight great-grandparents.

Twenty generations ago you have 1,048,576 great-great-....-great grandparents. So far, no problem.

Thirty generations ago you have over a billion (1,073,741,824) great-great-....-great grandparents.

Now we have a problem. Thirty generations ago (which is over 600 years ago) there were about a half billion people on earth.

Does that mean that you don't exist?

1	2	16	65,536
2	4	17	131,072
3	8	18	262,144
4	16	19	524,288
5	32	20	1,048,576
6	64	21	2,097,152
7	128	22	4,194,304
8	256	23	8,388,608
9	512	24	16,777,216
10	1,024	25	33,554,432
11	2,048	26	67,108,864
12	4,096	27	134,217,728
13	8,192	28	268,435,456
14	16,384	29	536,870,912
15	32,768	30	1,073,741,824

Chapter Seven
Jelly Beans and Sugar Weasels

It was five minutes to nine. There wasn't enough time for Fred to start the next algebra topic. He knew many fun math puzzles that he would sometimes insert at the end of the class hour.

He told the class, "Take out a piece of scratch paper, and we'll have a race. The first one to get the answer please raise your hand.

"A man dug a hole in his garden that was 2 feet wide, 4 feet long, and 3 feet deep. How much dirt is in the hole?"

8:55

Chris didn't even have to write anything down. "It was 24 feet since 2 times 4 is 8, and 8 times 3 is 24."

Fred said, "Not quite right. You have the wrong units."

Pat said, "It should be 24 *cubic* feet."

Fred said, "The units are right, but the number is wrong."

Some of the students tried 4 × 2 × 3 instead of 2 × 4 × 3, but they still got 24.

One student asked, "Is the hole in the shape of a box? Does it have square corners and straight lines?"

Fred said, "Yes."

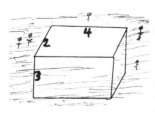

Terry liked to draw and asked, "Does it look like this?"

Fred said yes. He liked the pictures of the flowers that Terry drew around the hole.

It was nine o'clock. Fred said that if no one could correctly figure out how many cubic feet of dirt were in the

Chapter Seven   Jelly Beans and Sugar Weasels

hole, he would give the answer tomorrow at the beginning of the class.

Joe finally swallowed all the jelly beans he had jammed into his mouth. He said, "Mr. Teacher, I've got the answer."

All the other students snickered. In their experience Joe always had an answer, but it was never the right one. Today, they were mistaken.

Joe continued, "The way I see it, when you dig a hole, you take the dirt out. So that big hole that the guy dug doesn't have any dirt in it."

They all moaned. They knew he was right.

Fred dismissed the class and thanked Betty for taking over his class on such short notice.

Betty said that it was easy. She had studied two years of algebra, a year of geometry, and trigonometry in high school. In the first two years at KITTENS University she did calculus. She was now a first-year graduate student (after four years of getting her bachelor's degree as an undergraduate).

She knew the algebra, geometry, trig, and calculus cold. It was as simple as teaching the ABCs for her.

Joe came rushing back into the classroom. He said, "Joseev mezo dingaa sauwn osser!"

Joe spit all his sugar weasels into the garbage can. Now he could speak more clearly.

sugar weasel

small essay
Sugar Weasels?

Sugar weasels are the newest candy. They are made by the same people who produce Sluice, the world's sweetest soft drink.

Chapter Seven   Jelly Beans and Sugar Weasels

To make a sugar weasel they pour 8 ounces of Sluice into a pot. They heat it on a stove until it is boiled down to one-fourth of its original volume.

$$4 \overline{)\begin{array}{r} 2 \\ 8 \\ \underline{8} \\ 0 \end{array}}$$

There are now 2 ounces of a very thick, gummy mess in the pot. They scrape it out and form it into the shape of a weasel and then paint it with food coloring.

end of small essay

Joe always liked the purple sugar weasels because they turned his tongue purple. He thought that was funny.

Fred and Betty knew that Joe had something important to say. Joe does not normally spit out sugar. His usual route for sugar is into the mouth and down into the stomach.

Joe said, "You got to stay inside. Don't go outside."

"Why not?" Betty asked.

"There's a giant rhino out there. It looks just like my sugar rhino candy, but it's a zillion times bigger. It looks really mean. I called the university president and told him what I saw. He asked me what he should do. I told him to make sure that no one goes outside." Joe had said all this in one breath.

sugar rhino

Just then the tornado warning siren sounded.

Everyone in Kansas knows what that means. Everyone outside came inside. Everyone inside headed downstairs to the tornado shelter in the basement.

Chapter Seven   Jelly Beans and Sugar Weasels

The shelter was the safest place to be in case of a tornado. There were concrete walls, a solid door, and, of course, no windows.

They had to stay there until the all-clear siren sounded.

Two people were talking on their cell phones about what had just happened. Suddenly they realized that they were in the same room.

Joe opened his backpack and took out a 115-ounce bag of sugar weasels. In his head, Fred converted that 115 ounces into pounds. Since there are 16 ounces in a pound, he divided.

$$16\overline{)115} \quad \begin{array}{l} 7 \text{ R}3 \\ \underline{112} \\ 3 \end{array}$$

Divide: 16 goes into 115 seven times.
Multiply: 7 times 16 equals 112.
Subtract: 112 from 115.
The remainder is 3.

Fred thought That bag of sugar weasels weighs 7 pounds and 3 ounces.

The students and the teachers were used to tornado warnings. They had been in tornado shelters many times. It was one of the drawbacks of living in Kansas.

No matter where you choose to live in the United States (or in the whole world), there are major drawbacks.

☞ If you live in the South, there might be hurricanes.

☞ In the West, there might be earthquakes.

☞ In parts of the North, it can be very cold.

☞ In parts of the Southwest, there can be drought (lack of rainfall).

If there were a perfect spot that had no drawbacks, then a lot of people would move there, and it would become *very crowded*. And that would be a major drawback!

Chapter Seven — Jelly Beans and Sugar Weasels

Your Turn to Play

1. Four students in the shelter loved to play bridge. One of them shuffled the cards and dealt out the 52 cards to the four players. How many cards did each receive?

2. The sugar weasels in Joe's 115-ounce bag each weighed 5 ounces. How many weasels were in his bag?

3. It is difficult for most adults to watch Joe eat sugar weasels. It would be like watching someone eat spoonfuls of sugar. In addition, Joe chewed with his mouth open.

When Joe entertained himself by biting the heads off of the weasels, everyone moved away from him to the other side of the room.

Betty pulled her *Metamathematics* book out of her backpack. Today she was reading a proof that propositional logic is consistent. She found this fascinating—much more than playing cards or eating candy. She could read about two pages per hour. How many pages could she cover in ten hours of study?

4. Sometimes the all-clear siren would sound after only five or ten minutes. In those cases, the tornado had headed away from KITTENS instead of toward it.

Sometimes there was a series of tornados, and everyone would have to stay in the shelter for over an hour.

After 17 minutes had passed, Fred amused himself by converting that into seconds. Seventeen minutes equals how many seconds?

........COMPLETE SOLUTIONS........

1. Each student received one-fourth ($\frac{1}{4}$) of the cards.

```
      13
   4) 52
      4
      ‾‾
      12
      12
      ‾‾
       0
```

> Divide: 4 goes into 5 one time.
> Multiply: 1 times 4 equals 4.
> Subtract: 4 from 5.
> Bring down the 2.
>
> repeating these steps . . .
>
> Divide: 4 goes into 12 three times.
> Multiply: 3 times 4 equals 12.
> Subtract: 12 from 12.
> There is no remainder.

Each student would receive 13 cards.

2. How many 5-ounce weasels are there in a 115-ounce bag?

```
       23
   5) 115
      10
      ‾‾
       15
       15
       ‾‾
        0
```

> Divide: 5 goes into 11 two times.
> Multiply: 2 times 5 equals 10.
> Subtract: 10 from 11.
> Bring down the 5.
>
> repeating these steps . . .
>
> Divide: 5 goes into 15 three times.
> Multiply: 3 times 5 equals 15.
> Subtract: 15 from 15.
> There is no remainder.

There are 23 sugar weasels in a 115-ounce bag.

3. If Betty could read 2 pages per hour, she could read ten times as much in ten hours. $10 \times 2 = 20$ pages.

4. There are 60 seconds in a minute. Multiply or divide? We are expecting *more* seconds than minutes.

```
       60
     × 17
     ‾‾‾‾
      420
       60
     ‾‾‾‾
     1020
```

17 minutes = 1,020 seconds

Chapter Eight
Errors in Thinking

Three hours passed in the tornado shelter. It was about noon.* The bridge players got tired of playing bridge. Betty was just finishing up six pages of *Metamathematics*. Joe had consumed several pounds of sugar weasels.

Fred was thinking.

First thought: It has been very quiet outside. If tornados were passing by, they would be ripping buildings apart. It would be noisy.

Second thought: Joe told me, "I called the university president and told him what I saw. He asked me what he should do." This is really strange. If the university president had any brains, he would have asked where and when Joe had seen that giant rhino. Instead, he asked for advice . . . from Joe!

university president

Third thought: Joe had given the advice, "I told him to make sure that no one goes outside." Then we heard the tornado warning siren. Could those two events be connected?

Fred looked at Joe. His eyes were glazed over. He held a headless weasel in each hand. He had stopped eating.

---

* Some people say 12 noon. Is there any other kind of noon?

Joe stood up, walked quickly to the shelter room door, opened it, and headed outside.

He had gone outside to "make pizza." This happened to Joe very frequently after eating pounds of candy. Joe never seemed to learn.

After he had thrown up, he came back inside, leaving the door open. The sun shone into the room.

"The giant rhino is still out there," Joe said. "He's about as big as a car. I came back inside before he could attack me. And that giant rhino has two horns, not just one."

Chris asked him, "How did it grow an extra horn?"

Joe wasn't sure. "I may have miscounted. Anyway, I had to run away before it attacked me."

Fred walked to the door and looked outside.

small essay

The Biggest Error in Thinking

Just because two things happen together, does not mean that one of them causes the other.

First example: Just because Joe eats a ton of candy and thinks that a cow is a giant rhino does not mean that overeating candy causes an inability to tell a cow from a rhino. There is **a third thing that causes both**: namely, Joe isn't very smart.

Second example: Children who live in big houses tend to have straighter teeth. That's true, but does that mean that you should buy a bigger house so that your kids' teeth will be straighter? Because bigger houses and straighter teeth go together, that does not mean that one causes the other. There is **a third thing that causes both**: parents with more money tend to buy bigger houses and spend more for dental care for their kids.

Third example: Every couple of months there is an article in the newspaper with a headline like: MORE EDUCATION EQUALS HIGHER LIFETIME EARNINGS. There always is a table showing how much a high school graduate will make over a lifetime, and how much a college graduate will make. It is true that college graduates tend to make a lot more money than high school graduates. But that doesn't prove that more education causes more income. Just because two things happen together does not mean that one causes the other. There is **a third thing** that may promote both getting more education and getting more money: such as being smart. (Joe is going to have a real tough time graduating from KITTENS.)

high school graduate

college graduate

end of small essay

Fred's trig class was supposed to meet at noon. However, all his students were in the tornado shelters in all the buildings on campus. It would take an hour for Fred to run to all the shelters and announce that it was safe to come out.

The quickest way Fred thought would be to head to the president's house and have him sound the all-clear siren.

Chapter Eight   Errors in Thinking

Fred had never been to the president's house. It was located on the west side of the campus, far away from everything and everybody.

He started jogging west. He got to a dense forest that he had never seen. It got very dark as he jogged down the path through the trees.

He thought Forests like this do not occur naturally in Kansas. It must have cost a lot of money to have these big trees hauled in and planted.

The path came to a stone wall that must have been 30 feet tall. Ten of me Fred thought.

$$3 \overline{)30}^{10}$$

Fred is 36 inches, which is 3 feet, which is 1 yard. Thirty feet = 10 Freds = 10 yards.

Memorize this!

> In life,
>
> whenever you get to a wall,
>
> feel around.
>
> There will always be a door.

He didn't know whether to head to the left or to the right. Either way would eventually work.

Chapter Eight   Errors in Thinking

The only certain way to fail when you hit a wall is to stop.

Your Turn to Play

1. Fred headed to the left. After running 3,561 feet he came to a giant entrance gate. How many yards did he run?

2. If Fred had headed to the right instead of to the left, he would have had to run 4,672 feet.

What is the perimeter of this wall? (purr-RIM-eh-terr Perimeter = the distance around.)

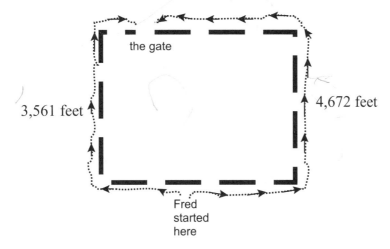

3. Copy on your paper and work it out: 96
                                        × 86

Chapter Eight   Errors in Thinking

........**COMPLETE SOLUTIONS**.......

1. In order to change 3,561 feet into yards, the big question is whether to multiply or divide. Since there are going to be fewer yards than feet, we divide.

```
       1187
    3) 3561
       3
       ‾
       5
       3
       ‾
       26
       24
       ‾
       21
       21
       ‾
```

3,561 feet equals 1,187 yards.

2. The distance around the entire wall would be the sum of 3,561 feet and 4,672 feet.

```
   3561
 + 4672
 ‾‾‾‾‾‾
   8233
```

The perimeter of the wall is 8,233 feet.

3.
```
     96
   × 86
   ‾‾‾‾
    576
    768
   ‾‾‾‾
   8256
```

Note from the author:

Did you ever notice that when I finish working out a problem in the Your Turn to Play, the last line is often an English sentence that answers the question?

In problem 2, the question was What is the perimeter of the wall? *My answer was* The perimeter of the wall is 8,233 feet. *I don't just scribble down some numbers.*

Chapter Nine
The President's House

Fred looked through the bars of the gate. He wanted to make sure that there wasn't a NO TRESPASSING sign, and he wanted to make sure there wasn't a big guard dog waiting inside to eat him.

Fred mentally created a "fred" function. Its domain was the set of all dogs, and the codomain was {nice dog, bad dog}. It looked like this:

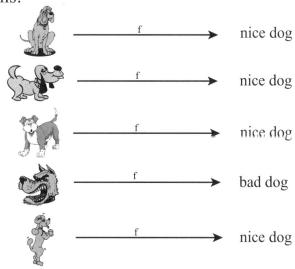

A function is not the domain or the codomain. It is the rule which associates to each member of the domain exactly one member of the codomain.

The little "f" over the arrow indicates that this is the fred function.

Someone else could make a different function using the same domain and codomain. Let's suppose that person is a grouch that hates all dogs. We could call his function "g."

Here is what the grouch's function would look like:

Chapter Nine   The President's House

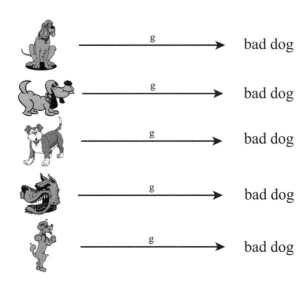

Grouch should probably never choose to become a veterinarian!

When Fred walked through the gate, he didn't see any NO TRESPASSING signs or any dogs. What he did see were acres and acres of beautiful green lawn. An acre is a measure of area like square inches or square feet.

Real estate agents who sell houses often list the size of a house in square feet and the size of land in acres. In the real estate classes, they are taught that an acre is equal to 43,560 square feet, but that is hard to imagine.

It is much easier to think of an acre as a square that is about 208 feet on each side.

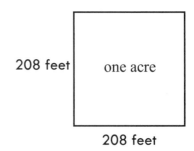

Chapter Nine   The President's House

In the distance Fred could see a very large mansion. It would be the perfect place for a husband and wife and a dozen kids.

Not counting his 52 servants (butlers, maids, cooks, gardeners, etc.), the president lived alone.

It took ten minutes for Fred to walk across the lawn to the front door. As he approached, spotlights turned on, a video camera focused on him, and the doorbell rang automatically.

Fred thought to himself This is a lot of security.

When a woman answered the door, Fred didn't know what to do. He felt very short.

She said nothing.

Is she a security guard? Fred wondered. She's not dressed like a maid. Am I in trouble? Does she think I'm here to apply as a boy-butler? Did I use the wrong entrance? I hope I don't get arrested. Why is she wearing sunglasses?

"I'd like to see the president, if I may," Fred said.

"Don't worry, kid," she said. "It's part of the mansion tour. I will be your tour guide. Five bucks, please, and then go

Chapter Nine The President's House

stand over with the rest of the crowd. We will start the tour very soon."

Mansion tour? Is that how he makes all his money? If there are 20 people in the tour, and each one pays $5, then he would make $100 for each tour.

```
 20
× 5
100
```

If they do 8 tours each day, then he would make $800 each day.

```
100
× 8
800
```

If they did tours six days each week, he would make $4,800 each week.

```
800
× 6
4800
```

With 52 weeks in a year, he would make $249,600 each year.

```
4800
× 52
9600
24000
249600
```

In 4 years, that would be $998,400, which is almost a million dollars ($1,000,000).

```
249600
×     4
998400
```

Fred checked his pockets to see if he had five dollars.

He found:

✓ An uneaten cracker in his back pocket. It had broken into six equal pieces when he sat on it last week. Each piece was one-sixth of the whole cracker.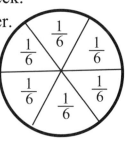

✓ Some string.

✓ Some fur from a tennis ball.

✓ His library card.

✓ Pieces of a jelly bean that he had found on the floor of his classroom.

Chapter Nine   The President's House

✓ His KITTENS faculty card.

✓ A scrap of paper—from his calculus lecture notes.

> $\lim_{x \to a} f(x) = L$ is true if . . .
> for every $\varepsilon > 0$, there exists a $\delta > 0$ such that
> $0 < |x - a| < \delta$ implies $|f(x) - L| < \varepsilon$.

(This is the definition of limit, which is the only really new concept in the two years of calculus. Everything else in calculus is based on this idea of limit. It usually takes students several weeks to start to feel comfortable with this definition.)

Your Turn to Play

1. The cracker in Fred's back pocket was broken into six equal pieces. Each piece was one-sixth of the whole cracker.

$$\frac{1}{6} + \frac{1}{6} + \frac{1}{6} + \frac{1}{6} + \frac{1}{6} + \frac{1}{6} = 1$$

How would the first three lines of this question look if the cracker had broken into four pieces instead of six?

2. If Fred made $39 for each lecture he gave,
    and he gave 7 lectures each day,
        for 5 days each week,
            for 48 weeks each year,
how much would he make in a year?

3. What is the perimeter of this property?

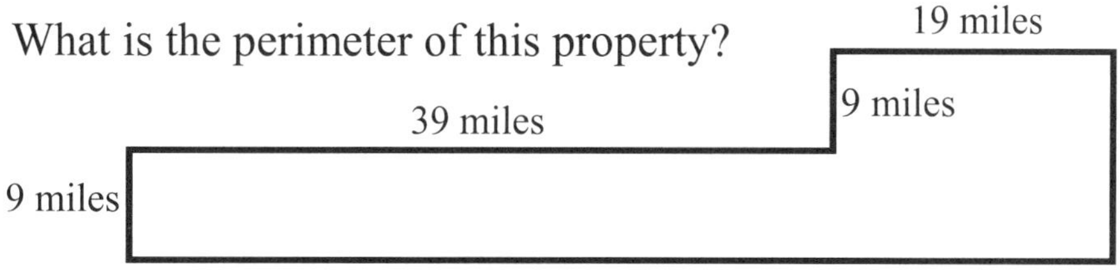

(Diagram: an L-shaped property with labels: 19 miles, 39 miles, 9 miles, 9 miles, 9 miles)

Chapter Nine The President's House

·······COMPLETE SOLUTIONS·······

1. If the cracker in Fred's back pocket was broken into four equal pieces, then each piece would be one-fourth of the whole cracker.

$$\frac{1}{4} + \frac{1}{4} + \frac{1}{4} + \frac{1}{4} = 1$$

2. If he gave 7 lectures each day and earned $39 for each lecture, he would make $273 each day.

```
  39
×  7
 273
```

```
 273
×  5
1365
```
If he earned $273 each day and he worked 5 days each week, he would earn $1,365 each week.

If he earned $1,365 each week and worked for 48 weeks each year, he would earn $65,520 each year.

```
  1365
×   48
 10920
  5460
 65520
```

3. The first step is to fill in the missing dimensions.

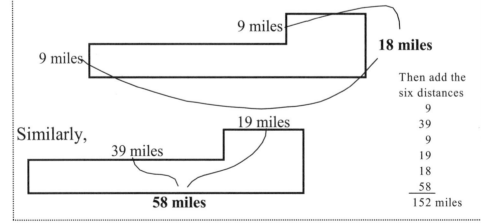

Then add the six distances
```
   9
  39
   9
  19
  18
  58
 152 miles
```

Chapter Ten
The Tour

The last thing in Fred's pocket was, of course, the $1,000 check that the police officer had given him as a reward for helping capture C. C. Coalback.

"I'm sorry," Fred said. "I don't have any cash. All I have is this check."

"That's no problem," the tour guide said. She pointed to the ATM.

Something seemed very strange, but Fred couldn't quite figure out what it was. Then he realized: *The university president has an ATM inside his home!*

Fred borrowed a chair and stood on it. He deposited his check and withdrew $60 in cash. Forty dollars was to pay the debt he owed Kingie and twenty dollars was so that he could pay the tour guide.*

Fred paid and then went to stand with the other 19 people who were going to tour the president's home.

A second thing seemed strange. Fred remembered that he wanted to see the president so that the all-clear siren would be sounded. Where did those 19 people come from?

Fred knew almost everyone at KITTENS, but he didn't recognize any of the people on the tour.

---

* The KITTENS Bank ATM only dispenses twenty-dollar bills. If it dispensed five-dollar bills, then Fred would have withdrawn only $45 —$40 for Kingie and $5 for the tour guide.

Chapter Ten The Tour

They were tourists! They had just gotten off the bus and had never heard the tornado warning siren.

The tour guide said, "Follow me. The first stop will be the award-winning Grand Kitchen."

When they got to the Grand Kitchen, everyone gasped. The stove had 24 burners. Fred didn't have to count them all. There were 4 rows of burners and 6 in each row.

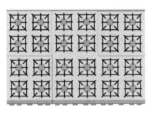

Fred took a peek into the pantry. He couldn't believe how much food was there. You could feed the entire KITTENS population for a week.

In the kitchen pantry

Fred counted six cooks working in the kitchen. One was cooking a turkey. One was making cream puffs. One was chopping vegetables. One was stirring soup. One was kneading bread. One was washing his hands.

Fred asked, "Isn't this a lot of food for just one person? How could the president eat this much for lunch?"

The tour guide had been asked this question many times. She said, "This allows the president to have a choice for lunch."

The woman with the camera asked, "And . . .?"

Chapter Ten   The Tour

"What he doesn't eat, we just throw away," the tour guide answered.

small essay
When Someone Else Pays

KITTENS University pays for the president's mansion, for his 52 servants, for his food, for everything.

It is a basic law of economics:

If it is free, you tend to waste it.

The president usually has a hot dog and a can of Sluice for lunch, but each day he makes the cooks prepare a feast.

You can't imagine how much toothpaste he puts on his toothbrush. Every time he wipes his hands on a towel, he throws it on the floor. Then one of his servants replaces it with a new towel. He leaves the television on when he leaves the room. A servant will turn it off for him. Dirty socks are left on the floor. If he spills cereal or milk, the maid will clean up after him.

end of small essay

"Our next stop," the tour guide said, "will be one of the smaller bathrooms."

All 21 people easily fit into that bathroom. (19 tourists + Fred + the tour guide) All 21 people could have easily fit into the shower.

It had three huge shower heads.

Chapter Ten   The Tour

The bathtub was so big that it needed a lifeguard.

The knobs on the sink were made of gold. Someone (you know who!) had left the water running and a bar of soap on the floor. Fred picked up the soap and turned off the water. He put the soap back on the sink. A maid came by, took the soap, and threw it in the trash. "The president never uses the same bar of soap twice."

The tour guide said, "Before we get to the master bedroom, are there any questions?"

"Yes. I have one," the woman with the camera said. "KITTENS University gives the president this mansion and the servants and a salary. What does the president do to earn all this?"

"He does two important things. He gives speeches, and he pushes the tornado warning siren button."*

"But why pay him so much?" the woman with the camera asked.

The tour guide didn't want to answer that question. She just said, "It's a long story.** We need to move on to the master bedroom."

The bed in the master bedroom wasn't twin size. It wasn't queen size. It wasn't king size. It was president size. It was so large that it took four maids just to straighten the sheets.

---

* It suddenly occurred to Fred that no one at the president's mansion had gone to a tornado shelter. That's because the siren sounds everywhere on the campus except inside his house. He doesn't like the sound.

** The truth is quite simple. The board of directors at KITTENS University sets the president's salary and benefits. Somehow, the president became head of the board of directors. Then the president would hold board of directors' meetings without telling the other members.

At these meetings, at which he was the only one present, he would vote on what his own salary and benefits would be.

Chapter Ten   The Tour

Your Turn to Play

1. A bed sheet for a president size bed measures 37 feet by 48 feet. What is the area of one of those sheets?

2. One of those sheets would be too large to fit in a washing machine. Instead, each day the maids would just throw out the old sheets and put brand-new ones on the bed.

   The pure silk pillowcases each cost $98. How much would pillowcases cost for the 17 pillows on his bed?

3. Fred had given the tour guide a $20 bill to pay for the five-dollar tour. How much change did he receive?

4. Suppose you had one-dollar, five-dollar, and ten-dollar bills in your cash register, and you wanted to pay $15.

   Could you do that with only one bill?

   Could you do that with exactly two bills?

   With exactly three bills?

   With exactly four bills?

   With exactly five bills?

   With exactly six bills?

   With exactly seven bills?

   With exactly eight bills?

   With exactly nine bills?

   With exactly ten bills?

   With exactly eleven bills?

   With exactly twelve bills?

   With exactly thirteen bills?

   With exactly fourteen bills?

   With exactly fifteen bills?

........COMPLETE SOLUTIONS........

1. The area of a rectangle is length times width.

$$\begin{array}{r} 37 \\ \times\ 48 \\ \hline 296 \\ 148 \\ \hline 1776 \end{array}$$

The area of a bed sheet is 1,776 square feet.

2. Seventeen pillowcases @ $98 each.
@ is the symbol for *at*.

$$\begin{array}{r} 98 \\ \times\ 17 \\ \hline 686 \\ 98 \\ \hline 1666 \end{array}$$

Seventeen pillowcases would cost $1,666.

3. $20 – $5 = $15

$$\begin{array}{r} \overset{1}{2}\!0 \\ -\ \ 5 \\ \hline 1\ 5 \end{array}$$

4. Since there is no such thing as a $15 bill, you couldn't pay with a single bill.

Two bills? Yes. $10 + $5

Three bills? Yes. $5 + $5 + $5

Four bills? Five bills? No.

Six bills? Yes. $10 + $1 + $1 + $1 + $1 + $1

Seven bills? Yes. $5 + $5 + $1 + $1 + $1 + $1 + $1

Eight bills? Nine bills? Ten bills? No.

Eleven bills? Yes. $5 + $1 + $1 + $1 + $1 + $1 + $1 + $1 + $1 + $1 + $1

Twelve bills? Thirteen bills? Fourteen bills? No.

Fifteen bills? Yes. $1 + $1 + $1 + $1 + $1 + $1 + $1 + $1 + $1 + $1 + $1 + $1 + $1 + $1 + $1

Chapter Eleven
The Rooms

The guide announced that the tour was almost over. The last thing was a chance to see the president. Fred got excited. It would be his opportunity to tell the president to sound the all-clear siren.

They walked down a long hallway with many doors on each side. One of them was marked Garden Room.

The woman with the camera opened the door and peeked in. The smell of all the flowers was overwhelming.

The Movie Room had a giant theater screen and one seat.

The Game Room was filled with video games, a pool table, and thousands of arcade games.

the Garden Room

The Snack Room had a popcorn machine, a pizza oven, a Sluice machine, an ice cream maker, a doughnut machine, and a score* of vending machines that dispensed cookies, crackers, and candy. None of the vending machines required money to operate. All you had to do was push the buttons.

The Lake Room was one of the larger rooms.

---

* A score is 20. A dozen is 12. A brace is two. If someday you have to go to the dentist to get your teeth straightened, you might receive a brace of braces.

Chapter Eleven The Rooms

At the end of the long hallway, the tour guide announced, "And this is our president." She pointed to an oil painting on the wall.

Some of the tourists glanced at the painting. One of them said, "Nice boat." None of them were eager to actually meet the president.

His mansion was interesting. Being rich doesn't make a person interesting.

The nineteen tourists left the building.

Fred turned to the guide and said, "I really need to see the president. It's university business."

She took off her sunglasses and looked at Fred. "I recognize you. You are Fred Gauss. I took advanced algebra from you last year. At first I thought you were just some little kid."

Fred didn't know what to say. He was a kid. He was little.

The tour guide continued, "The president only does two things: give speeches and push the tornado warning siren button. Since that button has already been pushed, the only university business would be if you wanted him to give a speech."

Chapter Eleven The Rooms

Fred had heard many of the president's speeches over the last five years. They were long and boring, and they all sounded alike. He wasn't here to ask for another speech.

"I'm here to ask him to sound the all-clear siren. There is no tornado danger. The president pushed the tornado warning button because he was told that there was a giant rhino running around on campus. He was misinformed. The giant rhino turned out to be a very gentle cow."

"The president is a very busy man. He has to give speeches at the beginning and end of the school year, and he has to push the tornado warning button. He can't be expected to perform other activities such as sounding the all-clear siren. Besides, he is not here right now. After he pushed the tornado warning button early this morning, he got on his private jet for a much-needed vacation to Anaheim, California."

A million thoughts went through Fred's head. He's very busy? He gives the same two speeches every year. I teach from 8 to 5 every day, except for the five-minute break from 3 to 3:05. Fred was much too polite to say these thoughts aloud. Instead, he asked, "How does the all-clear siren get sounded if the president doesn't do that?"

She smiled. "Easy. That's part of my job." She turned and pressed a button on the wall. "The president also mentioned that there will be a school holiday until he gets back from his vacation."

all-clear siren button

"When will he get back?"

"He'll be gone until graduation day, which is ten days from now. He has to give a speech at the graduation ceremony."

Fred couldn't believe it. "That means classes are canceled for the rest of the semester! My beginning algebra class will never learn about absolute values. My advanced algebra class will never learn about permutations. My trig class will never learn about the polar form of complex numbers."*

The tour guide shrugged her shoulders. "I guess those trig students will never learn about the polar bear form of complex numbers."

Fred explained, "It's the polar form, not polar bear. You were in my advanced algebra class. You learned that x^5 meant x times x times x times x times x. x^5 = xxxxx. So 2^5, for example, is equal to $2 \times 2 \times 2 \times 2 \times 2$, which is 32."

She asked, "What's that got to do with the polar form?"

"When you get to the last part of trig, you will solve the equation $x^5 = 32$."

"That's duck soup,**" she said. "Everyone knows that if x is equal to 2, then x^5 will equal 32."

Fred smiled. "But there are *four other numbers* which will make x^5 equal to 32. You can find them using the polar form of complex numbers and de Moivre's theorem."

She thought for a moment and said, "That's nuts. If you stick any number bigger than 2 into x^5, you will get an answer bigger than 32. For example $3^5 = 3 \times 3 \times 3 \times 3 \times 3$, which is 243. If you stick any number smaller than 2 into x^5, you will get an answer smaller than 32."

She took out her calculator and computed $(1.9)^5$, which is $1.9 \times 1.9 \times 1.9 \times 1.9 \times 1.9$, and got an answer of 24.76099.

---

* These are the last topics in each of those courses.

** *Duck soup* is an idiom. It has nothing to do with cooking ducks in water. Duck soup means something that is easy to do. The expression entered the English language somewhere between 1910 and 1915.

76

Chapter Eleven   The Rooms

She asked, "Are any of those other numbers that make $x^5 = 32$ true—are any of those other numbers larger than 2?"

"No. None of them are greater than 2."

"So those four other numbers are all less than 2?"

"No. And, of course, those other four numbers are not equal to 2," Fred answered.

"Wait a red-hot minute!" she exclaimed. "You can name a number that is not larger than 2, or smaller than 2, or equal to 2? Tell me! I need to know!"

Fred was silent for a moment. "Can you see why I am so sad that I won't be able to teach this last week of trig?"

Your Turn to Play

1. $1^5 = ?$
2. $0^6 = ?$
3. $6^2 = ?$
4. $10^3 = ?$
5. $1^{872,899,550,214} = ?$
6. How much does four score equal?
7. How much does five dozen equal?
8. Name one solution for $x^3 = 64$. (There are two other solutions, but you won't be able to name them until you have had the last week of trig.)
9. The local zoo wants to get rid of their extra polar bears. They advertised, "Seventeen polar bears @ $783 each." If you bought all of them and brought them home, would your parents be happy?

.... COMPLETE SOLUTIONS ....

1. $1^5 = 1 \times 1 \times 1 \times 1 \times 1 = 1$
2. $0^5 = 0 \times 0 \times 0 \times 0 \times 0 = 0$
3. $6^2 = 6 \times 6 = 36$
4. $10^3 = 10 \times 10 \times 10 = 1{,}000$
5. $1^{872,899,550,214} = 1$
6. Four score is four times 20, which is 80.
7. Five dozen is five times 12, which is 60.

$$\begin{array}{r} 12 \\ \times\ 5 \\ \hline 60 \end{array}$$

8. Two doesn't work since $2^3 = 8$.
   Three doesn't work since $3^3 = 27$.
   Four works since $4^3 = 4 \times 4 \times 4 = 64$.

9. I doubt it.

Before you tell your mother that you want to grow up to be a university president, you should be warned that not all university presidents are like the one at KITTENS University.

Most of them do not push tornado warning buttons. Most do not have a mansion with a Lake Room. Most do not have the power to set their own salaries. Most of them have other duties besides speech making, such as fund raising and representing their universities at ceremonies.

Search on the Internet under "salaries university presidents" if you are curious.

Chapter Twelve
Nine Days

Fred thanked the tour guide and walked outside. He didn't know what to do with himself now. Classes had been canceled for the rest of the semester. He loved teaching, but now he found that he couldn't teach for a while.

He had a new idea for teaching the polar form for complex numbers. He would get a big polar bear doll and bring it into his trig class. He was sure that his students would love that.

Fred's teaching always had lots of fun and surprises. He knew that students don't learn much when they are bored.

Right outside the entrance gate was a newspaper stand.

He looked at all the newspapers and magazines that were for sale. He wasn't interested in the bridal magazines. Or the martial arts magazines. Or the magazine that announced, "Lose 19 pounds in 19 days!" Fred weighed 37 pounds, and he did not want to lose 19 pounds.

Chapter Twelve   Nine Days

There was a newspaper in German and another one in French. Then he found what he was looking for, the campus newspaper. He paid the man a dime.

THE KITTEN Caboodle

The Official Campus Newspaper of KITTENS University   Monday Edition 10¢

news flash!

University Closed Till Graduation Day

KANSAS: The KITTENS University president announced this morning that all classes have been canceled for the rest of the semester.

President Takes a Vacation

He offered no reason for the closure of the campus.

advertisement
Nine-Day Bowling Ball Sale!
Buy now! Get a lifetime supply. A dozen for $132.

advertisement
Learn to sell vacuums like a professional.

advertisement
Nine Day Adventure!

Camp Horsey-Ducky offers . . .
★ *Life in the outdoors!*
★ *Horses!*
★ *Thrills!*

Contact: Miss Ente at 555-2691

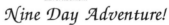

Chapter Twelve Nine Days

Fred knew exactly what he wanted to do for the next nine days. He borrowed the phone from the newspaper salesman and called Miss Ente.

Fred knew why the camp was called Horsey-Ducky. In German, *Ente* means duck.

Fred dialed the number. "Hello. This is Fred Gauss. I read about your Horsey-Ducky camp in the newspaper."

"This is Miss Ente. The camp starts this afternoon. We are going to have lots of fun. This is the perfect time to sign up."

That is all that Fred needed to hear. "Yes! Sign me up."

"Please meet me at the KITTENS bus station at a quarter to four. Bring along all the clothes and toiletries* that you will need for the nine days. The camp fee is $300. Please report to Dr. Morningstar at the KITTENS Hospital for your camp physical. We want to make sure that everyone who attends camp is healthy. I'll set up your appointment for a quarter after two."

Fred knew what "a quarter after two" meant. If you divide an hour into quarters, each quarter would be 15 minutes.

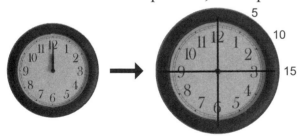

---

* Toiletries = all the things you use in the bathroom to groom and clean yourself, such as toothpaste, toothbrush, floss, comb, brush, soap. Really old people, such as teenagers, will also need deodorant and maybe shaving stuff.

Chapter Twelve   Nine Days

Fred didn't have a lot of time to get ready for his nine-day adventure. He raced back to his office, climbed the two flights of stairs, hurried down the hallway past the nine vending machines (four on one side and five on the other), and dashed into his office.

a quarter after two

He stuffed socks, underwear, shirts, pants, and toiletries into a suitcase. He added one bow tie just in case he needed to do some teaching.

He told Kingie that later today he would be heading off to Camp Horsey-Ducky for nine days of adventure.

Kingie was just finishing up an oil painting. It was one in his series of famous cowboys. Earlier in the day he had done "The Colorado Kid," "Montana Mike," and "Idaho Ike."

When Fred looked at the painting, he realized he didn't have a cowboy hat. He knew that he needed a hat that would make him look like a real cowboy.

"Reno Stan"
by Kingie

And he would need a handkerchief to wrap around his neck like the one in the oil painting. And rope for the cows. And spurs. And cowboy boots. And a harmonica to play around the campfire in the evening. And a book of cowboy songs. And a compass so that he wouldn't get lost on the prairie. And an ax to cut firewood. And a canteen because you could get mighty thirsty riding in the dust behind all those cows. And a dozen math books to read at night after the other campers had headed off to bed. And a lantern in order to read the math books. And extra batteries for the lantern. A book of cowboy jokes to entertain everybody around the campfire: "A TERMITE WALKED INTO A

Chapter Twelve   Nine Days

SALOON AND ASKED, 'WHERE'S THE BARTENDER?'" Fred thought that joke was so funny. He giggled and giggled.

Your Turn to Play

1. Fred needs to meet Miss Ente at the bus station at a quarter to four. Draw a picture of a clock that is set to a quarter to four.

2. The advertisement for bowling balls stated, "A dozen for $132." How much would each ball cost at that rate?

3. Fred weighs 37 pounds. If he lost 19 pounds, how much would he weigh?

4. 10^3 equals one thousand. How much does 10^6 equal?

5. The camp fee is $300 for nine days of adventure. Roughly, how much is that per day? Just give your answer to the nearest dollar.

6. Fred didn't have the time to go to the store to buy all the things he needed for camp. On the Internet he located a website. What would be the total cost?

Price List

Cowboy hat. Size extra-small	17.
Neck handkerchief	6.
Rope for cows	8.
Spurs, silver with gold trim	55.

Chapter Twelve   Nine Days

........COMPLETE SOLUTIONS.......

1.
3:45

2. A dozen bowling balls cost $132. If you are not sure whether to add, subtract, multiply, or divide, the General Rule is to restate the problem using simple numbers. Then look at what you would do with the simple numbers. For example, if 3 bowling balls cost $12, then each one would be $4. You divided.

So, in the original problem, you divide:
$$\begin{array}{r} 11 \\ 12\overline{)132} \\ \underline{12} \\ 12 \\ \underline{12} \end{array}$$

Each ball would cost $11.

3. $\begin{array}{r} 37 \\ -19 \\ \hline 18 \end{array}$ Fred would weigh 18 pounds.

4. $10^6 = 10 \times 10 \times 10 \times 10 \times 10 \times 10 = 1,000,000$
   10^6 is equal to one million.

5. Using the General Rule, we see that we should divide.
$$\begin{array}{r} 33 \\ 9\overline{)300} \\ \underline{27} \\ 30 \\ \underline{27} \\ 3 \end{array}$$ It would cost roughly $33 per day.

6. Adding up a big column of numbers is not easy the first time you do it. It does get easier with practice. The sum is $86.

Chapter Thirteen
To See Dr. Morningstar

On another website he found the rest of the items he needed. He already had the dozen vacation math books on his shelves, so he didn't need to buy them.

Boots for riding the range. Size 4	37.
Harmonica. Key of C	29.
Campfire songbook	8.
Compass, ax, canteen	52.
Lantern and spare batteries	45.

He added those numbers in his head and got an answer of $171. Please add those numbers and see if you can also get the answer that Fred got.

Fred looked at the clock. His medical appointment was at a quarter after two. He had ten minutes to get there.

He gave Kingie a hug and headed out the door.
    Past the 9 vending machines (4 + 5 = 9).
        Down two flights of stairs.
            Past the tennis courts.
                Past the university chapel.

At the rose garden, he stopped to smell the roses. The yellow roses were his favorite. He turned south toward the KITTENS Hospital. (Turning north would have taken him to the Administration Building.)

Chapter Thirteen To See Dr. Morningstar

At the hospital entrance was a listing of all the doctors and their room numbers. The list was alphabetized. Fred looked under the letter M:

    Dr. McClean   439
    Dr. MacDonald   506
    Dr. Morningstar   814

He thought to himself *They must be using the* ANGLO-AMERICAN CATALOGUING RULES VERSION 2 *for their alphabetizing. If I didn't know about those rules, I would have thought that they had made an error.*

In the way that offices are often numbered, Fred knew that Dr. McClean was on the 4th floor, Dr. MacDonald was on the 5th floor, and Dr. Morningstar was on the 8th floor.

He ran up the stairs instead of taking the elevator. Exercising made Fred feel good. As he headed up the stairs he wondered what Dr. Morningstar would look like. Would he be like this: Or like this:

> Truth: Your thoughts of the future are often much worse than what will really happen.

He entered room 814 and told the receptionist, "I'm Fred. I have an appointment at 2:15."

The receptionist smiled. "The doctor will see you right now." It was 2:15.*

---

* No paperwork to fill out? No waiting for ten or twenty minutes after the appointment time to be seen? No seeing the nurse before seeing the doctor? No request for insurance papers?

    There is a special word for this. It is called *fiction*.

Chapter Thirteen To See Dr. Morningstar

> Hi. I'm Mary Morningstar. Do you like to be called Fred or Freddie? I understand you are here for a physical exam.

She had Fred sit up on a table. She listened with her stethoscope.

First, she listened to his heart to make sure that it was beating okay.

Then she asked Fred to take some deep breaths. She was listening to his lungs to make sure that they were okay.

"Perfect!" she said. "The last thing we need is a urine specimen." She handed him a plastic cup and pointed down the hallway.

Here is where things got a little difficult. Fred didn't know what the word *urine* meant. The smart thing to do would have been to ask. Fred didn't. Instead, he just heard, ". . . we need a specimen."

One meaning of *specimen* is an example. Fred thought that Dr. Morningstar wanted a writing sample. He asked, "Does neatness count?"

She didn't know what to say and finally sputtered, "Just put it in the cup."

She headed back to the nurses' station. She couldn't stop laughing. She had heard many questions over the years when she had asked for a urine sample, but she had never been asked if neatness counted.

Chapter Thirteen — To See Dr. Morningstar

Fred headed down the hallway and found a chair. He took out a piece of paper and wrote:

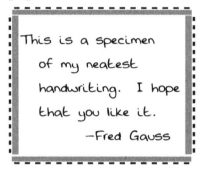

This is a specimen of my neatest handwriting. I hope that you like it.
—Fred Gauss

He rolled up the paper and put it in the plastic cup. He brought it back to Dr. Morningstar and handed it to her. She read the note and realized that Fred didn't know what a urine sample was.

She handed him another cup and told him what he needed to do. Fred headed down the hallway to the restroom and filled the cup.

As he headed back to the doctor, he thought *Urine samples are a lot nicer than giving blood samples. I like this doctor.*

"That completes the exam, Freddie," Dr. Morningstar said. "Do you have any questions?"

Fred realized that he had never answered her question about whether he liked Fred or Freddie. The doctor seemed friendly, and so Fred asked her, "What's the urine specimen for?"

She smiled. "We test for a lot of things." She showed him a chart:

Color	☺	pale to dark yellow
	☹	red (blood in the urine)

Chapter Thirteen   To See Dr. Morningstar

Odor	☺	slightly like the smell of nuts (If you have been eating asparagus, it will smell different, but that is normal.)
	☹	bad odor may indicate infection
Ketones	☺	none
	☹	sometimes found if the patient has not eaten for 18 hours or longer
Glucose (sugar)	☺	none
	☹	may be caused by diabetes, an adrenal gland problem, liver damage, brain injury, or pregnancy

Your Turn to Play

1. Can you guess which one of the four parts of Fred's urine test (color, odor, ketones, and glucose) will be ☹?

2. Which floor of the hospital will Dr. Mueller (room 765) be on?

3. If you had an appointment at a quarter past two and you had to wait in the waiting room until half past two, how many minutes would you have waited?

4. Fred noticed that Dr. Morningstar had a box of cotton swabs. It was labeled: Contains 17 dozen swabs. How many swabs is that?

5. If she sees 7 patients each hour and works for 8 hours each day, 5 days each week, 49 weeks each year, how many patients would she see in a year?

Chapter Thirteen — To See Dr. Morningstar

> ....COMPLETE SOLUTIONS.......
>
> 1. Who knows when the last time Fred has eaten anything? He will most probably have ketones in his urine.
> 2. Room <u>7</u>65 is on the 7th floor.
> 3. A quarter past two is 2:15. Half past two is 2:30. You would have waited fifteen minutes.
> 4. A dozen is 12. Seventeen dozen is 17 times 12.
>
> ```
>       17
>      ×12
>       34
>       17
>      204
> ```
> Seventeen dozen is equal to 204.
>
> 5. Eight hours @ 7 patients/hour equals 56 patients/day.
>
> Fifty-six patients/day @ 5 days/week equals 280 patients/week.
>
> Two hundred eighty patients/week @ 49 weeks/year equals 13,720 patients/year.
>
> ```
>      280
>    ×  49
>     2520
>     1120
>    13720
> ```

Except for Fred, most people don't understand what is funny about: A TERMITE WALKED INTO A SALOON AND ASKED, "WHERE'S THE BARTENDER?" (From Chapter 12)

Question: What do termites eat? Answer: Wood.
Question: What are bars often made of? Answer: Wood.
WHERE IS THE BAR . . . TENDER?

What is funny to five-year-olds like Fred is often not very funny to older people.

Chapter Fourteen
Leisure

It was a quarter to three. Fred had to meet Miss Ente at a quarter to four at the bus station. He had an hour. All the things that he had ordered on the Internet would be delivered to the bus station. He had his checkbook to pay the camp fee. The station was nearby.

In short, Fred had an hour of free time. From the hospital he walked back to the rose garden. Leisure* time is as important as work time.

Everybody knows what you do during work time. You work.

The list of what can be done in leisure time is long:
nap
  shop
    eat pizza
      read
        go jogging
          play cards
            watch television
              play tag
                play the piano

---

* Leisure = free from working. The word *leisure* does not obey the spelling rule: *I before E except after C.*
    It doesn't even obey the longer rule:
        I before E
        Except after C,
        Or when sounded as A
        As in neighbor or weigh.

Chapter Fourteen Leisure

But there is one leisure time activity that is more important than any of these, even more important than eating pizza!

If you are working as a carpenter, you think about carpentry. You keep your mind on your work—or you can get hurt!

In contrast, leisure gives you time to think, to ask the bigger questions in life. Many of the most important advances in the history of the world happened when people spent their leisure time thinking about:

What makes plants grow?
How do stars shine?
Why is it cold in winter?
When will people stop doing bad things?

Fred was standing in the rose garden. It was leisure time. He was thinking . . . about his medical exam, about giving a urine specimen. He knew that he had a bladder—a "bag" inside of his body that held the urine. He knew that when his bladder was full, it was time to head to the bathroom. Everyone knows this.

But Fred went one step further. He thought But how does my bladder get filled? It doesn't happen by magic. Somehow the urine gets into the bladder. How?

Many people have never thought of that. Fred did. He was curious about lots of things.

He knew it would be silly to ask the roses, "How does my bladder get filled?"

He couldn't ask his parents since they weren't around.

His doll, Kingie, couldn't tell him. Kingie didn't have a bladder.

Fred ran back to his office. Running at 8 feet per second, he was back to the Math Building in 79 seconds. Back in his office, he pulled Prof. Eldwood's *Parts of the Body* off the shelf.

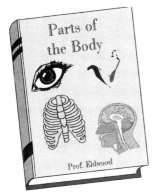

He looked in the index for *bladder* and found that it was discussed on pages 522–528. He took the book to his desk, opened to page 522, and started reading. He wanted to know how urine gets into the bladder.

Prof. Eldwood first described where the bladder is located in the body. Then its size. Then he mentioned the duct (the tube) that carries urine *from* the bladder to the outside world—the urethra (you-WREATH-rah).

Finally, on page 528 he talks about the ducts (the tubes) that carry urine *to* the bladder. Those ducts are called ureters (your-REET-ters).*

But what are the ureters connected to? Fred wondered.

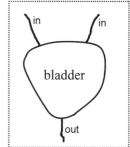

In the last sentence of the section on bladders, Prof. Eldwood said that the ureters connect the kidneys to the bladder.

KIDNEYS! Fred flipped back to the index and looked up *kidney*. He was surprised. Bladder had been covered on pages 522–528. Kidneys had a much longer discussion—over

---

* Those words are close to each other: urethra and ureters. The way to remember which is which is that urethra rhymes with underneath-ra.

a hundred pages long. Bladders just hold a pool of urine. Kidneys really *do* something.

    The book answered a lot of Fred's questions:

Question: Why are there two ureters?

Answer: Because there are two kidneys.

Question: Why are there *two* kidneys?

Answer: You have two eyes, two ears, two lungs, two halves of your brain, two hands—two is a very popular number for important parts of your body.

Question: But I have only one heart.

Answer: There are two parts to your heart. They are known as left and right. One part pumps blood to the lungs, and the other half pumps blood to the rest of your body.

Question: Where are kidneys located?

Answer: On either side, just under the bottom of your ribs. If you get tackled in football from the side or the back, your kidney can be injured. A mild tackle can result in pain in your lower back and traces of blood in your urine. A big blow to your kidney can cause kidney failure.

Question: What if I lose both kidneys? Am I dead?

Answer: No. There are machines that can do what kidneys do. They are called dialysis machines. The other possibility is a kidney transplant—putting someone else's kidney into your body.

Question: What do kidneys do—other than just make urine?

Answer: Their real job is to clean your blood.

Question: Clean my blood! I didn't know that it was dirty.

Answer: Do you remember the story of Goldilocks and the Three Bears? When she ate the bears' porridge, one bowl was too cold, one was too hot, and one was just right. The same is true for everything in your blood. The kidneys make sure that the concentrations of nitrogen, water, glucose, acid, etc., are

Chapter Fourteen   Leisure

just right. If the concentration of a chemical gets too high, the kidneys filter it out and send it to the bladder. Your kidneys are really smart.

Speaking of really smart, it is now . . .

Your Turn to Play

1. To find one-half of something, you divide by 2. Half of 100 is 50.

$$2\overline{)100} \quad \frac{50}{100}$$

$\frac{1}{2}$ of 100 is 50.

To find one-third of something, you divide by 3. One-third of 18 is 6.

$$3\overline{)18} \quad \frac{6}{18}$$

$\frac{1}{3}$ of 18 is 6.

One common size for a hamburger is a quarter pound. What is one-fourth of a pound in ounces? (One pound = 16 ounces.)

2. A couple of pages ago, Fred ran back to his office from the rose garden. He ran for 79 seconds at the rate of 8 feet per second. How far did he run?

3. In *Parts of the Body,* bladders were discussed on pages 522–528. How many pages was that? (This answer isn't 6!)

4. The word *leisure* is not the only word in English that does not obey the rule: *I before E*

   Except after C.

   Or when sounded as A

   As in neighbor or weigh.

It is weird. Without looking in foreign dictionaries, neither German nor French, can you seize upon another word that doesn't obey the rule? Your fame will be raised to a great height if you can think of another word, but you will not forfeit any points if you can't.

Chapter Fourteen Leisure

........COMPLETE SOLUTIONS.......

1. Four ounces. $4\overline{)16}$ gives 4

The reason we asked how many ounces in a quarter pound hamburger is so that you could make a comparison. An adult kidney weighs about 5 ounces. Many adults don't know this. Often they will guess, "a pound."

2. 79 seconds at the rate of 8 feet per second. If you are not sure whether to add, subtract, multiply, or divide, the General Rule is to restate the problem using simple numbers. Then look at what you would do with the simple numbers. If you ran at the rate of 2 feet per second for 3 seconds, you would go 6 feet. You multiplied.

So we multiply the rate times the time to get the distance.

$$\begin{array}{r} 79 \\ \times\ 8 \\ \hline 632 \end{array}$$ Fred ran 632 feet.

3. Let's look at a simpler question: *How many pages would you read if you read pages 1–8?* The answer is 8 pages. If you subtract 1 from 8, you are finding out how far 8 is from 1.

If you are on the 1st floor and you walk up to the 8th floor, you will climb 7 flights of stairs.

If you read pages 522–528, you have read 7 pages.

4. Let me give you a hint. I wrote:

It is *weird*. Without looking in *foreign* dictionaries, *neither* German nor French, can you *seize* upon another word that doesn't obey the rule? Your fame will be raised to a great *height* if you can think of another word, but you will not *forfeit* any points if you can't.

Chapter Fifteen
Packing

Fred put his *Parts of the Body* book back on the shelf. It went right next to Prof. Eldwood's *Parts of a Car* book. That book is only 178 pages long. Cars are much less complicated than bodies.

In case you are interested, here are other *Parts* books that Fred owns:

Parts of Denver (a travel guide)

Parts of Speech (nouns, pronouns, verbs, etc.)

Parts in Hair (standard part: wide part:)

Fred picked up his suitcase that he had packed with socks, underwear, shirts, pants, toiletries and a bow tie. Fred did not use a full-sized suitcase.

There was no easy way for him to lift it.

Instead, he had packed everything into his lunch box. His extra-small socks, extra-small underwear, extra-small shirts and pants all fit easily into it. He had several lunch boxes and picked the one with a duck on it. He thought Miss Ente would like that since *Ente* in German means duck.

Chapter Fifteen   Packing

He gave Kingie another hug and said, "I'll see you in nine days."

He headed down the hallway, down the two flights of stairs, across the campus quad, and by the KITTENS University library. He suddenly realized that he had forgotten to pack any math books. He had been thinking about packing them, but it had slipped his mind.

Fred thought to himself *Imagine going on a vacation and not bringing along any math books!*

He headed into the library. He had plenty of time before a quarter to four, which is when he was supposed to meet Miss Ente at the bus station.

Fred knew exactly where the math books were located. He had been there many times. The math section was divided into many categories:

set theory $\{1, 2, 3\} \cup \{3, 4\} = \{1, 2, 3, 4\}$

logic   "Roses smell nice" is the same as "If it doesn't smell nice, it isn't a rose."

algebra $5x^2y + 2x^2y = 7x^2y$

geometry

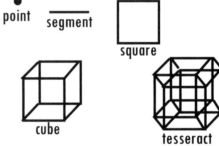

trig $\sin^2 x + \cos^2 x = 1$

statistics   Mean, median, and mode averages

topology   A square is the same as a circle.

metamathematics   Not everything that is true can be proved.

Someone once counted 128 subfields in all of mathematics.

Chapter Fifteen Packing

Fred wanted to pick out a dozen set theory math books to take with him on his nine-day adventure. Twelve books he checked out of the library dealt with:

General set theory
Kripke-Platek set theory
Tarski, Mostowski, and Robinson S' set theory
Zermelo set theory
Ackermann set theory
Zermelo-Fraenkel set theory
Von Neumann-Bernays-Gödel set theory
Morse–Kelley set theory
Tarski–Grothendieck set theory
New Foundations set theory
Scott-Potter set theory
Positive set theory

He also checked out several algebra books.

Outside, he began reading about Zermelo set theory and balanced the rest of the books on his head.

This was a very happy Fred.

Here is how Fred would Read & Walk: He would read several sentences and then walk a little while thinking about what he had just read. Then he would stop and read a little bit more.

Doing two things at the same time (called multi-tasking) is a bad idea. A couple of years ago, Fred used to read while he was walking. He found out about multi-tasking when he walked into a fire hydrant. His nose hurt for a week.

The human brain was not designed to do two things at the same time. For example, trying to study math and listen to the radio at the same time means that you will do both poorly.

Chapter Fifteen Packing

Fred dreamed that after he read those dozen books, he might invent Fred Gauss set theory. He wondered what it might look like. Everyone else in set theory puts the members of a set in braces like this: {cow, ☎, ⊛}.

Maybe Fred Gauss set theory could list the members of a set vertically.

{
cow
☎
⊛
}

Or maybe Fred Gauss set theory could list the members of a set all on top of each other. { c⊛ẘ☎ }.

> Truth: You often have to think of a lot of ideas that aren't very good before you find one that is really great.

small essay
How Math Happens

Where does mathematics come from? Did you ever think about that?

Do you think that people look in aquariums and see fish blowing TWO PLUS TWO EQUALS FOUR? Here is a secret: Fish don't invent math (and most don't blow bubbles).

Do you think that people just walk outside and read math off of rocks? Here is a secret: Rocks aren't good at math.

Do you think that mathematics has always just existed and that it is just the same as it always has been?

Chapter Fifteen Packing

Here is a secret: More mathematics has been invented in the last twenty years than in the four thousand years before that!

People create math. Human beings sit down and devise new mathematics. They are doing that today, and they are doing it faster than any other time in history.

You cannot choose a better time in history to become a mathematician.

end of small essay

Your Turn to Play

A little set theory:
1. The union of two sets is everything that is in either set or in both sets.
    {house, moon} ∪ {house, star} = ?
    {1, 2, 3} ∪ {7, 8} = ?
2. The intersection of two sets is everything that is in both sets.
    {house, moon} ∩ {house, star} = ?
    {1, 2, 3} ∩ {7, 8} = ?

A little logic:
3. Does "If it is a pizza, then I like it" mean that "If I don't like it, then it is not a pizza"?
4. Does "If it is a pizza, then I like it" imply that "If I like it, then it must be a pizza"?

A little exponents
5. Ten thousand equals 10 with what exponent? ($10^?$)
6. 87^2 = ? (multiply it out)

........COMPLETE SOLUTIONS........

1.   {house, moon} ∪ {house, star} = {house, moon, star}
One of the "spelling rules" for listing the members of a set is that you do not repeat a member.

   {house, h~~ouse~~, moon, star}

> The reason for this rule is that we want to make it easy to count the number of members in a set. If there are duplicates in the listing it makes it harder to count.
> For example, it would be hard to count the (unique) members in this set:
> {δ, ξ, λ, ζ, δ, γ, ξ, ζ, δ}.

   {1, 2, 3} ∪ {7, 8} = {1, 2, 3, 7, 8}

The order in which you list the members of a set does not matter. {1, 2, 3, 7, 8} = {1, 3, 2, 8, 7} = {8, 7, 3, 2, 1}

2.   {house, moon} ∩ {house, star} = {house}
   {1, 2, 3} ∩ {7, 8} = { }

{ } is called the **empty set**.

3. "If it is a pizza, then I like it" and "If I don't like it, then it is not a pizza" are called **logically equivalent**—they both say the same thing.

   In the symbols of logic, which you will study in geometry, P → Q and not-Q → not-P are logically equivalent.

4. "If it is a pizza, then I like it" does *not* imply "If I like it, then it must be a pizza." I also like beef ribs, cheesecake, boysenberry pie, and chocolate fudge ice cream.

   In the symbols of logic, P → Q does not mean that Q → P.

5. Ten thousand = 10,000 = 10^4

6. 87^2 = 87 × 87 = 7,569

```
   87
  ×87
  ───
  609
  696
  ────
 7569
```

Chapter Sixteen
Spurs

Fred finished the Zermelo set theory book and began the Ackermann set theory book. Fred's power of concentration allowed him to get a lot more done than those who divide their attention.

He suddenly noticed that he hadn't been walking at all, just reading and thinking about what he had been reading.

3:40

He had five minutes before he was supposed to meet Miss Ente. He closed the Ackermann book and hurried toward the bus station.

The previous time he had been to the bus station was when he took a trip to Edgewood. This time the station looked a little different. All of the things he had ordered on the Internet had arrived. There were 13 boxes. Four of them were stacked in front of the building and nine of them were behind it.

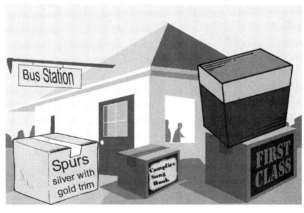

Fred opened the box with the size 4 boots and tried them on. They were just a little loose. Next he opened the box with the spurs in it. He wanted to show Miss Ente that he was serious about going to Camp Horsey-Ducky.

103

Chapter Sixteen   Spurs

He was not used to walking while wearing spurs. They jangled with each step he took. He took them off and put them back into the box.

He thought I don't want to wear them out. I'll wait until I get out on the range. In reality, Fred wasn't sure what spurs were used for.

It was 3:45.

"Hi. Are you Fred?"

He carefully put the tape back on the box

Fred turned and knew immediately that this was Miss Ente. He had it all figured out.* Here she was with a horse. Her name was Ente, which is duck in German. And she was riding a horse. Fred thought And that is why she called the camp Horsey-Ducky Camp. "Yes, I'm Fred. I'm all ready for nine days of adventure."

He pointed to all the boxes, to his math books, to his shoes. Miss Ente seemed pleased. She kept on smiling. And smiling. And smiling.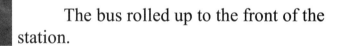

The bus rolled up to the front of the station.

---

* This is an example of adumbration. Adumbration comes from the Latin word *adumbrare*, which means to shade or overshadow.

Adumbration means to foreshadow or give a hint as to what is coming next.

If you go to a play and there is a shotgun leaning up against the wall in the first act, that adumbrates that it will probably be used in a later act.

When Fred "has something all figured out," you can guess that he will be wildly wrong.

Chapter Sixteen Spurs

Fred turned to pick up his boxes, and Miss Ente said, "My goodness. You are really prepared for anything."

Fred turned to thank her, and she said, "Just load the boxes into the bus." What was very strange was that she kept talking but didn't move her lips. Her face remained frozen in a happy smile.

But there was one thing that Fred was not prepared for. We will find out after this *Your Turn to Play*.

Your Turn to Play

1. Fred had started out with 13 boxes. He had emptied the box with the boots in it, so he had 12 boxes to load onto the bus.

The bus driver told him that the first box would cost $2.
The second box would cost $5.
The third box would cost $8.
Each box would cost $3 more than the previous box.

Fill in the blanks: $2 , $5 , $8 , $, $, $, $, $, $, $, $, $35 .

2. 2, 5, 8, . . . is called an arithmetic (air-rith-MED-ic) sequence (SEE-quence). The common difference (d) between any two consecutive terms is the same.

In 2, 5, 8, . . . the difference is 3. ($d = 3$)

What is the common difference in the arithmetic sequence 7, 16, 25, 34, 43, . . . ?

3. Arithmetic sequences are studied in the second year of high school algebra (known as advanced algebra or algebra 2). Years from now, when you study advanced algebra, we will call the first term of the sequence a, the last term l, and the number of terms n. We will prove that $l = a + (n-1)d$.

What is the 101st term of 6, 10, 14, 18, 22, . . . ?   (Hint: n = 101)

Chapter Sixteen   Spurs

.... COMPLETE SOLUTIONS ....

1. $2 , $5 , $8 , $11 , $14 , $17 , $20 ,
    $23 , $26 , $29 , $32 , $35 .

2. The common difference in the arithmetic sequence 7, 16, 25, 34, 43, ... is 9. $d = 9$

   The difference between 16 and 7 is 9. $16 - 7 = 9$
   The difference between 25 and 16 is 9. $25 - 16 = 9$
   The difference between 34 and 25 is 9. $34 - 25 = 9$

3. We want the 101st term of 6, 10, 14, 18, 22, ... so the number of terms equals 101. $n = 101$

   The first term is 6. $a = 6$
   The common difference is 4. $d = 4$
   The formula for the last term l is $l = a + (n - 1)d$.

   Therefore, $l = 6 + (101 - 1)4$
   $= 6 + (100)4$
   $= 6 + 400$
   $= 406$ The 101st term is 406.

After *Life of Fred: Fractions, Life of Fred: Decimals and Percents,* after the pre-algebra books, after *Life of Fred: Beginning Algebra,* will come *Life of Fred: Advanced Algebra.*

On page 281, in *Life of Fred: Advanced Algebra*, in problem 2, you will be asked: "What is the 200th term of 7, 9, 11, 13, ... ?"

This part of advanced algebra you can do right now.
The first term a is 7. The common difference d is 2. And n is 200.
$l = a + (n - 1)d = 7 + (200 - 1)2 = 7 + (199)2 = 7 + 398 = 405$. Easy!

Chapter Seventeen
Miss Ente

Fred loaded the 12 boxes into the bus. He owed the driver 2 + 5 + 8 + 11 + 14 + 17 + 20 + 23 + 26 + 29 + 32 + 35 dollars. This was an arithmetic series.

2, 5, 8, 11, 14, 17, 20, 23, 26, 29, 32, 35   an arithmetic sequence

2+5+8+11+14+17+20+23+26+29+32+35   an arithmetic series

The first term in a series is also called a.
The last term in a series is also called l.
The common difference in a series is also called d.
The number of terms in a series is also called n.
We use the same letters for both the sequence and the series.

Fred needed to know the sum (which is called s) of the arithmetic series 2 + 5 + 8 + 11 + 14 + 17 + 20 + 23 + 26 + 29 + 32 + 35 so that he could pay the bus driver. He paid him $222 without adding up the numbers!

When KITTENS University wasn't on vacation, Fred taught advanced algebra from 10 to 11 every day. He knew the formula for the sum of an arithmetic series:

$$s = (½)n(a + l)$$

(Multiplying by ½ is the same as dividing by 2.)

Fred had 12 boxes, so $n = 12$.
The first term in the series was 2, so $a = 2$.
The last term in the series was 35, so $l = 35$.

So the sum was $s = (½)(12)(2 + 35)$
$= (½)(12)(37)$
$= 6(37)$
$= 222$

"How did you add up all those numbers so fast?" Miss Ente asked. "You must be a math genius."

Fred was about to tell her about $s = (½)n(a + l)$ when he realized that  didn't move her lips when she talked and she didn't blink.

My name is Miss Ente.

Fred fainted. He hit the ground so hard that his boots came off.

In a couple of seconds, Fred opened his eyes.

Miss Ente explained to him: It's okay. A lot of people have this reaction to my talking. I'm glad you didn't beat me like Balaam did (Numbers 22:28). The world is full of strange and wonderful things.

A couple of weeks ago I read about a five-year-old who has been teaching math at KITTENS University for years. Now I find that really unbelievable.

Fred didn't know what to say. He knew about parrots who can say "Polly want a cracker," but those parrots don't understand what they are saying. This horse did understand.

Do you have any questions? A lot of people do when they learn I can think and talk.

Fred put his boots back on, thought for a moment, and asked a ton of questions.

Chapter Seventeen   Miss Ente

Question: Who or what is riding on your back?
Answer: Her name is Dolly. She is my doll. Boys and girls have dolls. So do I. She doesn't say anything because she is just a doll.

Fred thought to himself My doll, Kingie, can talk. If I tell Miss Ente about my doll, she probably won't believe me.

Question: Horses usually have names like Trigger or Buttermilk or Silver. I have never heard of a horse called *Miss*.
Answer: That's easy. I'm not married. It's tough to find a male horse that can say, "I do."

Before Fred could ask another question, the bus driver told him to get on the bus. "We have a schedule to keep, and I don't want to be late."

Fred asked Miss Ente, "Are you getting on the bus?"

She said That would be silly.* Horses don't ride on buses. I'll just gallop along behind the bus.

Fred picked up all his math books and his lunch box suitcase and got on the bus.

He went to the back of the bus so that he could see Miss Ente galloping behind the bus. As he sat down, his boots fell off. They were a bit too large for his feet. He checked the label inside the boot.

> Infant size 4 ✹ Euro 19 ✹ UK 3 ✹ Inches 4.5" ✹ cm 11.4

"Euro" stands for the measurement used in Europe.
"UK" stands for United Kingdom.

---

* Silly? Nothing in the *Life of Fred* books is ever silly. And I, your author, am a 600-pound pink hummingbird.

Time Out!

Many adult Americans have heard of the **United Kingdom** and can say that it has "something to do with England," but don't know more than this.

Here are the facts: The United Kingdom consists of four countries: England, Scotland, Wales, and Northern Ireland.

(Most of Ireland is independent and does not belong to the United Kingdom.)

Many adult Americans have heard of **Great Britain** but are not sure how that is different than the United Kingdom.

Great Britain = England, Scotland, and Wales.

Your Turn to Play

1. Using sets, Great Britain = {England, Scotland, Wales}.
{England, Scotland, Wales} ∪ {Northern Ireland} = ?

2. Boots in infant size 4 measure 4.5 inches. That means four and one-half inches.

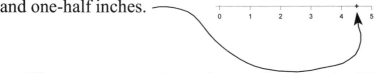

What can you say about the size (in inches) of Fred's feet given the fact that size 4 is too large for him?

3. Why is this *not* an arithmetic sequence?
    4, 6, 9, 11, 13

4. Why is this *not* an arithmetic sequence?
    4 + 6 + 8 + 10 + 12 + 14

5. There are five letters used in studying arithmetic sequences and arithmetic series: *a, l, d, n,* and *s.*
   Rearrange those letters to make a word.

   > *a* is the first term
   > *l* is the last term
   > *d* is the common difference
   > *n* is the number of terms
   > *s* is the sum of the terms

6. What is the 67th term of
    8, 14, 20, 26, 32. . . .

7. What is the sum of the first 67 terms of
    8 + 14 + 20 + 26 + 32 +. . . .

.... COMPLETE SOLUTIONS ....

1. {England, Scotland, Wales} ∪ {Northern Ireland} =
   {England, Scotland, Wales, Northern Ireland}
   which is also known as the United Kingdom

2. If size 4 is too large for him, Fred's feet must be less than 4.5 inches. In symbols, Fred's feet < 4.5".

3. 4, 6, 9, 11, 13 is not an arithmetic sequence because the difference between any two consecutive numbers is not the same. The difference between 4 and 6 is 2. The difference between 6 and 9 is 3.

4. 4 + 6 + 8 + 10 + 12 + 14 is not an arithmetic sequence. It is an arithmetic *series*. Commas separate the terms of a sequence.

5. *a, l, d, n,* and *s* can be rearranged into *lands*.

6. To find the 67th term of 8, 14, 20, 26, 32. . . we first note that it is an arithmetic sequence.

   $a = 8, d = 6, n = 67$ and we want to find l.
   The formula for the last term l is $l = a + (n - 1)d$.

   $l = 8 + (67 - 1)6$
   $ = 8 + (66)6$
   $ = 8 + 396$
   $ = 404$ The 67th term is 404.

7. To find the sum of the first 67 terms of 8 + 14 + 20 + 26 + 32 +. . . we first note that this is an arithmetic series with

   $a = 8, d = 6, n = 67$, and $l = 404$. (We found the value of l in the previous problem.)
   The formula for the sum (s) is $s = (½)n(a + l)$.

   $s = (½)(67)(8 + 404)$
   $ = (½)(67)(412)$
   $ = (½)(27,604) = 13,802$ _(Multiplying by ½ is the same as dividing by 2.)

Chapter Eighteen
Stacking Books

Fred had to do some quick thinking. He didn't want to arrive at Camp Horsey-Ducky and walk around in bare feet. His boots were too large. He thought of several possibilities:

☼ If I had a computer, I could order some size 3 boots and have them sent to Camp Horsey-Ducky.

> Drawbacks: I don't have a computer, and I don't know where the camp is located.

☼ I could glue my feet into the boots and then they wouldn't fall off.

> Drawback: I could never take my boots off.

☼ I could wait a year or two until my feet grow larger.

> Drawbacks: I haven't grown at all in the last couple of years, and I want to go to camp right now.

Fred thought of a fourth possibility. He went to the front of the bus and said to the bus driver, "Hello. I have a small question. Do you have any toilet paper?"

"Listen, kid," the bus driver began. "You should have thought about that before you got on the bus. Wait another couple of minutes. There's a gas station right up ahead."

Fred sat down in a seat near the driver and waited.

When they got to the gas station, the driver told Fred to hurry up. He had a schedule to keep.

Fred hopped off the bus, ran to the restroom, grabbed a handful of toilet paper, stuffed it into his boots, put on his boots, and ran back to the bus. Total elapsed time: 13 seconds.

His boots fit perfectly now.

Chapter Eighteen Stacking Books

The bus driver didn't know what to think. "Potty stops" (as some bus drivers call them) usually take about ten minutes.

Fred got back on the bus and thanked the driver. He said, "I have a small question."

"WHAT!" The driver was clearly exasperated.

"How far is it to Camp Horsey-Ducky?"

"I've never heard of it! Please kid. The map of the bus route is on the wall."

Fred studied the maps.

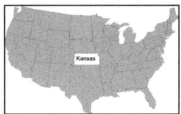

Bus Route

He couldn't see Camp Horsey-Ducky anywhere on the bus route. That map only showed cities and rivers.

He knew that there was no reason to worry.* He was sure that Miss Ente would tell him when to get off the bus.

He walked to the back of the bus. The bus driver was relieved that Fred was heading to the other end of the bus.

When Fred got to the back, he arranged the dozen set theory books into neat stacks. Fred didn't like to see messes. First, he put them into 4 stacks of 3 books.

---

* Do you remember what adumbration means?

Chapter Eighteen   Stacking Books

Then he changed them to 2 stacks of 6 books.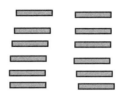

 Then 3 stacks of 4 books.

Then 6 stacks of 2 books, 1 stack of 12 books, 12 stacks of 1 book.

Twelve is such a good number. There are tons of combinations:
$12 = 1 \times 12 = 2 \times 6 = 3 \times 4 = 4 \times 3 = 6 \times 2 = 12 \times 1$

If he had 13 set theory books, there would be very few ways to stack the books.

He looked out the back window. He wanted to wave to Miss Ente. He knew that she couldn't wave back because she was using all four of her legs to gallop.

She wasn't there.

Maybe she is running alongside the bus he thought.

He checked both sides. No Miss Ente.

He ran to the front of the bus. He looked. She wasn't galloping in front of the bus.

She couldn't be on top of the bus.

Fred turned to the bus driver, "Excuse me. I have a big question."

The driver stiffened. He didn't know what to expect.

Fred asked, "Where is Miss Ente?" This was the third annoying question that Fred had asked.

"Who in blazes is Miss Ente? I have never heard of her."

Fred carefully explained, "She owns Camp Horsey-Ducky. [Not a good way to start. The bus driver had never heard of that camp.] She has a doll that has long hair, a white shirt, and likes to ride on Miss Ente's back. [Does this help explain who Miss Ente is?] Her doll cannot talk, but my doll can. Miss Ente can talk, and she knows what she is saying. She is not like a parrot who just repeats sounds."

small essay

Four Ways to Talk

Talking is more than saying words. When you were about one year old, you said your first word. You pointed to the family pet and said, "Dog." Your parents were very happy.

"Dog"

The first reason to talk is to **entertain**. You were very entertaining when you said your first word.

The second reason to talk is to **ask questions**. Your first sentence was, "Dad, may I borrow the car for the weekend?" Your father laughed. You were 14 months old.

The third reason to talk is to **convince**. "Dad, I really need the car and your credit card. There's a sale on at Tommy's Terrific Toys." Your father laughed. You were 15 months old.

The fourth reason to talk is to **inform**. "Dad, I have misplaced my pacifier." Your father did not laugh. You were six years old.

end of small essay

When Fred was teaching at KITTENS, he used all four ways of speaking with his students. He would tell funny stories

Chapter Eighteen   Stacking Books

about George and Cheryl Mittens and their three daughters, Fredrika, Medrika, and Paprika.* (**entertain**)

He would **inform**. He explained how a snake in Fredrika's dream that had eaten all the watermelons in her garden would lead directly to the Fundamental Theorem of Calculus.**

He would **ask** interesting questions (not just, "What is seven times eight?") and he would **convince** his students that math can be wonderful *if it is taught correctly.*

Your Turn to Play

1. Fred spent 3 seconds running from the bus to the restroom and 3 seconds running back to the bus. If his total potty stop time was 13 seconds, how long did he spend in the restroom stuffing toilet paper into his boots?

2. If Fred had 18 books, how many ways could he stack them? All the stacks will have the same number of books. Fred likes things to be neat. (One of your answers will be 2 stacks of 9 books.)

3. Find the sum of the first million terms of 43 + 44 + 45 + 46 + 47 + 48 +. . . .

(This is a two-step problem. First, find the last term: *l*.)

---

* In the tenth chapter of *Life of Fred: Calculus*.

** This was also in Chapter 10. And in case you are wondering, the Fundamental Theorem of Calculus is

$$\int_{x=a}^{b} f(x)dx = F(b) - F(a) \text{ where } F' = f \quad \text{... whatever that means!}$$

Chapter Eighteen   Stacking Books

......COMPLETE SOLUTIONS.......

1. If the total trip was 13 seconds and 6 seconds were used in running to and from the bus, then 7 seconds were spent in the restroom.

2.
    1 stack of 18 1×18
    2 stacks of 9 2×9
    3 stacks of 6 3×6
    6 stacks of 3 6×3
    9 stacks of 2 9×2
    18 stacks of 1 18×1

There are six ways he could stack the books.

3. $43 + 44 + 45 + 46 + 47 + 48 + \ldots$ is an arithmetic series. (That's the first thing to check. The common difference is the same between each pair.)

$a = 43$, $d = 1$, $n = 1{,}000{,}000$

The formula for the last term l is $l = a + (n - 1)d$.

$$\begin{aligned} l &= 43 + (1{,}000{,}000 - 1)1 \\ &= 43 + (999{,}999)1 \\ &= 43 + 999{,}999 \\ &= 1{,}000{,}042 \end{aligned}$$

The formula for the sum (s) is $s = (½)n(a + l)$.

$$\begin{aligned} s &= (½)(1{,}000{,}000)(43 + 1{,}000{,}042) \\ &= (½)(1{,}000{,}000)(1{,}000{,}085) \\ &= (500{,}000)(1{,}000{,}085) \\ &= 500{,}042{,}500{,}000 \end{aligned}$$

The sum is 500,042,500,000. This approach is a lot quicker than adding up $43 + 44 + 45 + 46 + 47 + 48 + \ldots + 1{,}000{,}042$.

Chapter Nineteen
Discovering Something New

Fred had never really answered the bus driver's question of who is Miss Ente. He began by saying that she owned Camp Horsey-Ducky and she owned a doll. *You are never defined by the things you own.**

Fred finally told the driver that Miss Ente had been galloping behind the bus.

The bus driver said, "When we were at the gas station for your potty stop, that horse and rider galloped past us. The rider said, 'I'll meet Fred at the next bus stop. That is where Camp Horsey-Ducky is.' That was funny. The woman talked without moving her lips."

Fred decided not to tell him it wasn't the woman who was talking. The bus driver seemed to be having a bad day, and Fred didn't want to annoy him by telling him he was mistaken.

On the other hand, this was great news for Fred. Fred thought Hurray! I haven't lost Miss Ente. She will be waiting for me at the next stop.

Fred did a little happy dance as he headed to the back of the bus. He passed a glass filled with polka dots that had been left on one of the seats. It had been there for months.

---

* If someone just wants to tell you all the things he or she owns, find someone else to talk with.

"I own nine Rag-A-Fluffy dolls. I own six horses. I own three houses. I own a giant computer. I own lots of clothes. I own."

Chapter Nineteen — Discovering Something New

When he got to his seat, he saw his stacks of set theory books. He also noticed the several algebra books that he had also checked out from the library. He opened an advanced algebra book and found the section on sigma notation. Σ is a capital S in Greek, and S was chosen to remind you of **sum**.

The easiest example was $\sum_{i=1}^{6} i$ which equals $1 + 2 + 3 + 4 + 5 + 6$.

$\sum_{i=1}^{6} i$ equals 21.

Fred suddenly discovered a connection that he had never thought of before: sigma notation and arithmetic series.

$\sum_{i=1}^{6} i = 1 + 2 + 3 + 4 + 5 + 6$ is an arithmetic series.

Time Out!

Many advanced algebra books never notice this connection. They teach about Σ in one chapter and arithmetic series in another chapter.

Many great discoveries in math have been made by connecting things that had never been joined before.

For example, in the 1600s, Descartes (day-CART) combined points from geometry with numbers from algebra.

He invented a whole new field of math called analytic geometry.

In the beginning of analytic geometry, the ordered pair of numbers (5, 2) is

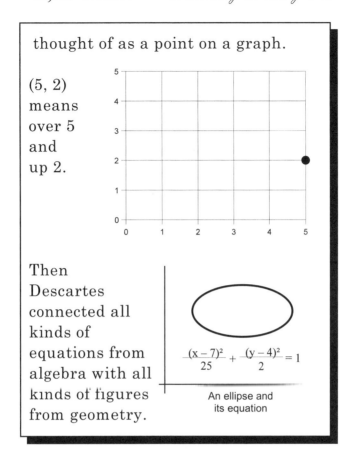

thought of as a point on a graph.

(5, 2) means over 5 and up 2.

Then Descartes connected all kinds of equations from algebra with all kinds of figures from geometry.

$$\frac{(x-7)^2}{25} + \frac{(y-4)^2}{2} = 1$$

An ellipse and its equation

Fred realized that really hard problems such as $\sum_{i=1}^{637} i$ could be solved in a matter of seconds rather than hours without having to add up $1 + 2 + 3 + \ldots + 635 + 636 + 637$.

Fred ran back to the glass that was filled with polka dots. He poured the dots out on the seat and arranged them.

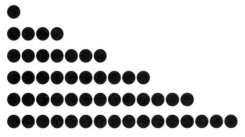

Chapter Nineteen   Discovering Something New

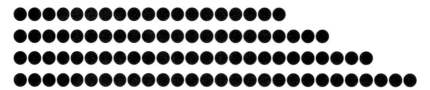

He had 10 rows. Each row had three more dots than the previous row. $a = 1$, $n = 10$, and $d = 3$. Fred was one of the few people in the world who knew how many polka dots can fit in that glass.

He scooped up the dots, put them back into the glass, and put the glass back where he had found it.

Fred looked out the window and spotted the Kansas Blood Bank. It was ten minutes after four.

The bus trip had started at a quarter to four (3:45).

The sun shone* on the dome of the blood bank. If this were December, the sun would be close to setting. It was the first day of June. The sun was still high in the sky.

Each day from Christmas to about June 21st the sun gets higher and higher in the sky. That is why it gets warmer as summer approaches.

On about June 21st the sun stops getting higher and higher. It seems to stand still on that day. In Latin, the word

---

* The past tense of *shine* is *shone* or sometimes *shined*. When it comes to polishing shoes, you say, "Yesterday, he shined his shoes."

  English is constantly changing. They used to call it electronic mail; then E-mail; then e-mail; and now it seems that email is the most common usage.

Chapter Nineteen   Discovering Something New

sol means sun, and the word *sistere* means to cause something to stand or to stop or put an end to.

Sol and *sistere* together give us our word *solstice*. The sun was approaching the summer solstice. The winter solstice, when the sun stops getting lower and lower in the sky, is on about December 22nd.

This had been a busy first day of June for Fred. The day wasn't over yet.

Your Turn to Play

1. Draw a graph and plot the points (1, 3) and (4, 2).

2. $\sum_{i=1}^{5?} i = ?$ (Work your answer out to a single number.)

3. How many polka dots can fit in that glass? (See the top of the previous page.)

4. The bus trip started at a quarter to four. It was now 4:10. How long had Fred been riding?

.... COMPLETE SOLUTIONS ....

1.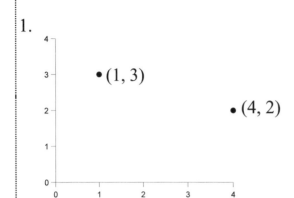

2. $\sum_{i=1}^{52} i = 1 + 2 + 3 + \ldots + 50 + 51 + 52$, which is an arithmetic series with $a = 1$, $d = 1$, $l = 52$, and $n = 52$. The sum (s) is $s = (½)n(a + l) = (½)(52)(1 + 52)$
$= (½)(52)(53)$
$= (26)(53)$
$= 1{,}378$

```
    53
  × 26
   318
   106
  1378
```

3. We want $1 + 4 + 7 + \ldots$ where $n = 10$. This is a two-step problem. First, we find the last term, l, and then we find the sum, s.

$l = a + (n - 1)d = 1 + (9)(3) = 28$

$s = (½)n(a + l) = (½)(10)(1 + 28) = 5(29) = 145$

Now you are one of the few people in the world that knows that 145 polka dots will fill a glass.

4. 25 minutes.

There are 15 minutes between 3:45 and 4:00.

There are 10 minutes between 4:00 and 4:10.

But only 24 minutes, 47 seconds if you count the 13 second bathroom stop. (page 113)

Index

$1,000,000 in four years... 64
100%... 38
12 noon... 55
2%... 31
absolute values... 75
adding two-digit numbers
.... 83-85
adumbration... 104, 114
alphabetizing can be tricky
.... 30, 36
analytic geometry... 120
Anglo-American Cataloguing
    Rules Version 2...
.... 30, 36, 86
area... 71
arithmetic sequence...
.... 105-107, 111, 124
arithmetic series... 107
Balaam... 108
basic law of economics... 69
being a good brother... 37
biggest error in thinking...
.... 56, 57
bladder... 93
brace... 73
calculus definition of limit
.... 65
centimeter... 18
Charles Demuth... 14

check writing (how to)...
.... 23, 29
  writing numerals... 24
converting feet into yards... 60
converting minutes to seconds
.... 54
converting ounces to pounds
.... 52
cowboy jokes... 82
decimal point... 14, 41
Descartes... 120
division with a remainder... 52
does more education imply
    more income?... 57
does neatness count?... 87
domain... 43, 44, 47
eager vs. anxious... 41
ellipse... 121
empty set... 102
every place has a drawback
.... 52
exponential equation... 76, 77
exponents... 77, 78, 83, 101
fiction... 86
find one-half of something
.... 95
find one-third of something
.... 95
fractions... 14, 64-66

function. 47
    codomain. 45, 47
    domain. 47
    image. 45
function 61, 62
    definition. 43
    domain. 44
Fundamental Theorem of
    Calculus. 117
g-r-a-y and g-r-e-y. 32
General Rule when you don't
    know whether
to add, subtract, multiply,
    or divide . . . .
    . . . . 24, 42, 84, 96
graduate student. 40
grams 41
graphing points. . . . .
    . . . . 121, 123, 124
Great Britain. 110
Great Depression 39
Guess-A-Function game. . . . .
    . . . . 45, 46
how many great math
    discoveries are made
    120
how not to carry money
    when walking on the street
    22
how to make boots fit. 113
idioms. 26, 28, 76
intersection of two sets. . . 101
job of a university. 19

Kansas map. 114
kidney. 93-95
kitchen pantry (world's
    largest?). 68
Lewis Waterman. 39
lifetime supply of bowling
    balls. 80
lingonberry jam. 28, 32
logic. 101
logically equivalent statements
    102
metamathematics 53, 98
milliliter 18
multi-tasking. 99
never write "alot". 32
numerals 14
one leisure time activity that is
    more important. . . . 92
one of the hardest parts of
    being a police officer
    34
parts of speech. 97
percent. 31
perimeter 59, 65
point, segment, square, cube,
    tesseract. 98
polar bear doll. 79
polar form of complex
    numbers. 76
quarter after two. 81, 82
reading clocks. . . . .
    . . . . 40-42, 49, 83, 84
results of urine tests. . . . 88, 89

score 73, 77, 78
semibreve 26
shoe sizes—US, UK, and Europe 109
sigma notation 120, 121, 124
sine function 98
small essays
   "Four Ways to Talk" . . . 116
   "How Math Happens" 100, 101
   "Sugar Weasels" 50
   "The Biggest Error in Thinking" 56, 57
   "When Someone Else Pays" 69
solstice 123
stethoscope 87
subtraction with borrowing 16, 21, 30
sugar weasels 50, 51
thirty generations ago 48
three keys to success 13
toiletries 81
union of two sets . . . 101, 111, 112
United Kingdom 109, 110
university presidents' duties 78
vacation without math books? 98
volume 49

whenever you get to a wall in life 58
whole, half, quarter, and eighth notes 26, 31
Why Being a Librarian Can Be Hard 30

If you would like to
learn more about
the books written
about Fred . . .

FredGauss.com